VOLUME 2

Laboratory Workbook & Program Documentation

CALCULUS CONNECTIONS
A *Multimedia* Adventure

Douglas Quinney • Robert Harding • IntelliPro, Inc.

JOHN WILEY & SONS, INC.
New York • Chichester • Brisbane • Toronto • Singapore

DOCUMENTATION

*CALCULUS CONNECTIONS
A MULTIMEDIA ADVENTURE*

Douglas Quinney
Robert Harding

Copyright 1996 John Wiley & Sons, Inc.
Portions Copyright 1994 Asymetrix Corporation
Portions Copyright 1994 Microsoft Corporation
Portions Copyright 1995 IntelliPro, Inc.
All Rights Reserved.

Copyright ©1997 by John Wiley & Sons, Inc.

All rights reserved.

Reproduction or translation of any part of this work beyond that permitted by Sections 107 and 108 of the 1976 United States Copyright Act without the permission of the copyright owner is unlawful. Requests for permission or further information should be addressed to the Permissions Department, John Wiley & Sons, Inc.

ISBN 0-471-13797-9

Printed in the United States of America

10 9 8 7 6 5 4 3 2 1

Printed and bound by Malloy Lithographing, Inc.

Authors / Developer

About the Calculus Connections Authors

Dr. Douglas Quinney, University of Keele, UK and Dr. Robert Harding, Cambridge University, UK, have written, developed and published multimedia mathematics software since the mid 1980's. Their groundbreaking works in multimedia mathematics include The Renaissance Project, commissioned by Apple Computer UK published in 1992 to demonstrate effectiveness of interactive multimedia technology in higher education, and major contributions to The United Kingdom Mathematics Courseware Consortium that includes over 20 UK Mathematics departments concurrently creating interactive multimedia modules for various mathematics courses.

About the Calculus Connections Developer

IntelliPro, Inc. is a leading developer of interactive courseware in mathematics, engineering and the sciences. It brings to the project extensive expertise in multimedia, mathematical simulation, computer graphics and animation, and instructional system design to construct this interactive multimedia environment for teaching and learning calculus.

Table of Contents

Laboratory Workbook

Preface

1 **Chapter 9**
Definite Integrals

21 **Chapter 10**
Rectilinear Motion

35 **Chapter 11**
Simpson's Rule

51 **Chapter 12**
Sequences and Series

71 **Chapter 13**
Differential Equations

95 **Chapter 14**
Spherical and Polar Coordinates

113 **Chapter 15**
Parametric Equations

131 **Chapter 16**
Mathematical Modeling

Program Documentation

153 **System Requirements**

153 **Installation Instructions**

153 **Navigation**

154 **Navigational Icons and Tools**

154 **Other Icons and Tools**
Hotwords
Help
Options
Mathematical Tools
References
Exit

155 **Technical Assistance**

155 **Volumes 1 & 3**

Preface

Workbook Overview

This Laboratory Workbook is a key component of the Calculus Connections multimedia teaching and learning experience. One workbook chapter accompanies each software module and is designed to:

- Provide background information
- Extend ideas introduced in the software
- Test understanding of concepts
- Apply this understanding to new situations

Connecting with the Software

We have designed each Workbook chapter to enhance your experience and maximize your potential for success with the corresponding software module. Chapters begin with necessary prerequisites, mathematical objectives, and a discussion of how the concepts you explore in each module relate to applications from engineering, physics, the sciences and daily life in general.

The Workbook uses your experience with the software as a springboard to explore other examples, exercises and new situations. It encourages multiple pathways for investigation, via approaches that call for pencil and paper, for computers, and for expressing your mathematical ideas and discoveries in sentence form. The pages in this book are perforated so that you can record your results on the Problem worksheets, tear them out, and hand them in to your instructor.

The Multimedia Experience

Our goal for Calculus Connections is to spark your imagination, deepen your mathematical intuition and give you opportunities to explore connections between mathematical theory and the world around you. We encourage you to question results, to construct "what if" scenarios and to experiment by changing parameters, inputting your own functions and generating your own graphics. Most of all, we encourage you to be curious, to be open-minded, and to have fun with this multimedia adventure!

D. Quinney
R. Harding

Chapter 9

Definite Integrals

Prerequisites

Before you study this material, you should be familiar with:

(1) The elements of differential and integral calculus (Modules 3, 5, 6, and 7).
(2) The concept that an integral can be thought of as "the area under the curve" (Module 6).
(3) The relationship between differentiation and integration, as expressed through the Fundamental Theorem of Calculus (Module 7).

Objectives

Many physical laws are given in terms of the rate of change of the variables involved with respect to time or other quantities. To understand and test a theory, it is important to find the values of variables predicted by these laws. This process requires that we go from knowing the function's rate of change to finding the function itself; it is the opposite of the process of differentiation in which we go from knowing a function to finding its rate of change.

In Chapter 7 the Fundamental Theorem of Calculus stated that integration and differentiation are inverse processes. Therefore, predicting variables from laws giving rates of change will involve integration. However, the process of finding an antiderivative introduces an arbitrary constant, and to take account of this in applications requires *definite integration*. As we have also seen (in Chapter 6 in particular), integration may also be thought of as a method of obtaining the area under a curve, and definite integration corresponds to specifying exactly which section of a curve to use. This module will describe some techniques that will help us apply definite integration to a number of commonly occurring mathematical expressions, either because we need to find areas or because we are interested in the predictions of laws involving rates of change.

Connections

Applications of integration can arise in the mathematics of business and management as well as in every branch of science and technology. Statistical techniques are very useful in all these areas, and the first connection between theory and application illustrated in Module 9 is in queuing theory, which helps to describe situations where a stream of events must be managed.

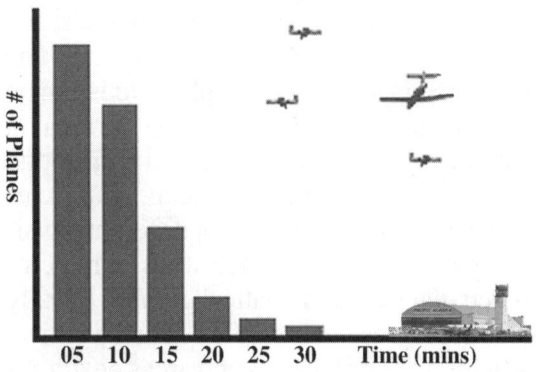

Figure 9.1. Histogram of airport plane queuing times.

The first video and application is about the arrival of aircraft at an airport, but it is easy to think of many other situations where queuing needs to be managed: for example, phone calls arriving at an exchange, customers arriving at a checkout, vehicles on a stretch of highway. A histogram can be used to describe the way that waiting times are distributed (see Figure 9.1), and it will be seen that mathematics can predict the shape of the envelope to such a histogram. The area under this curve can be used to answer questions such as "how many planes have to wait more than 10 minutes to land?", or "what proportion of planes have to wait from 10 to 20 minutes?"

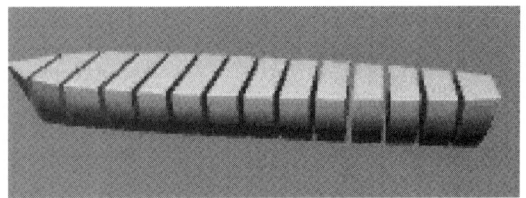

Figure 9.2. Volume of a sailboat hull.

The second video and application is the calculation of the volume of the hull of a sailboat (see Figure 9.2), assuming that the way its cross-sectional area changes along the length of the boat is known. It may at first seem strange that it is area that has been talked about in connection with integrals, and yet here it is volume that is being found. The explanation is that on a graph, an axis can be calibrated in units of area, so that an area on a mathematical graph can indeed correspond to a physical volume. In general, graph axes can correspond to all sorts of different physical quantities, and it must always be remembered that the mathematical graph is usually just the representation of an actual physical situation. Thus the axes and areas on graphs can correspond to any physical units we please, besides lengths and areas.

Definite Integrals and Area

This section repeats material from Chapter 6, where the concept of upper and lower sums was used to obtain the formulas:

$$\int_0^x mx\, dx = \tfrac{1}{2}mx^2 \qquad (9.1)$$

$$\int_0^x ax^2\, dx = \tfrac{1}{3}ax^3. \qquad (9.2)$$

(See Worked Examples 6.1 and 6.2.)

Recall that definite integrals may be interpreted as areas, provided that area refers to the mathematical area between a graph and the graph axis, in which regions where the graph is below the axis count as negative.

The integral $\int_0^x mx\, dx$ is the area under the line $y = mx$. The area can easily be found because it is the area of a right-angled triangle which is $\tfrac{1}{2}xy = \tfrac{1}{2}mx^2$. This is the most direct way to get this result, although it is reassuring that the method of upper and lower sums gives the same answer, as was shown in Chapter 6. Note that the derivative of $\tfrac{1}{2}mx^2$ with respect to x is mx, thus confirming in this case that differentiation is the inverse process to integration.

Note that any possible confusion between x used as the integration variable and x used in the limits can be avoided by writing, for example:

$$\int_0^x at^2\, dt = \tfrac{1}{3}ax^3. \qquad (9.3)$$

Now consider a more general case.

Exercise 9.1

Area under a line

Show geometrically that the area under the line $y = mx+a$, enclosed between the line itself and the interval $[p, x]$, is $\frac{1}{2}(x-p)(y(x)+y(p))$ (see Figure 9.3).

Show also that this expression is equal to $\left(\frac{1}{2}mx^2 + ax\right) - \left(\frac{1}{2}mp^2 + ap\right)$.

Hence find a function $F(x)$ such that
$$\int_p^x (mt+a)\,dt = F(x) - F(p).$$

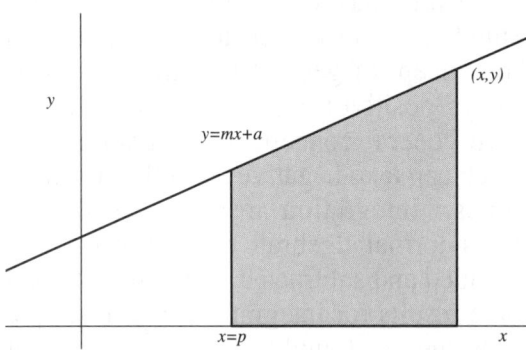

Figure 9.3. The area under y=mx+a on the interval [p, x].

Notice from the result of Exercise 9.1 that $F'(x) = mx+a$. Remind yourself from Chapter 7 that statements like

$$\int_a^b f(x)\,dx = F(b) - F(a) \quad (9.4)$$

may also be written as

$$\int_a^b f(x)\,dx = [F(x)]_a^b \quad (9.5)$$

In Worked Example 6.2 the integral of $f(x) = x^2$ was obtained using upper and lower sums. Here, consider a slightly different approach as illustrated in the software in the numerical methods concept (see Figure 9.4).

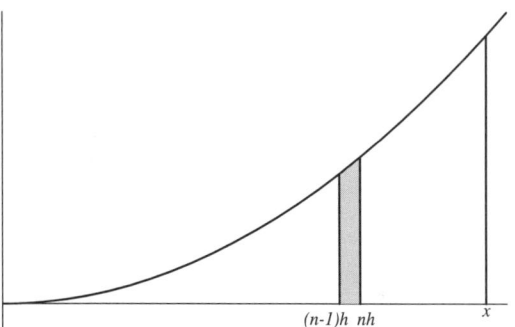

Figure 9.4. The area under y=x² on the interval [nh, (n+1)h].

Example 9.1

By dividing the interval $[0, x]$ into N subintervals of width h, and approximating the area on each subinterval with a trapezoid (see Figure 9.4), show that
$$\int_0^x t^2\,dt = \tfrac{1}{3}x^3.$$

Taking $f(t) = t^2$, the nth subinterval is $[(n-1)h, nh]$, and the results of Exercise 9.1 give the area of the nth trapezoid as:

$$\tfrac{1}{2}h\left[((n-1)h)^2 + (nh)^2\right].$$

The total area is therefore

$$\sum_{n=1}^N \tfrac{1}{2}h\left[((n-1)h)^2 + (nh)^2\right].$$

Using computer algebra or pencil and paper, we can show this to be equal to $\tfrac{1}{6}h^3 N + \tfrac{1}{3}h^3 N^3$. Since $x = Nh$, this equals $\tfrac{1}{6}h^2 x + \tfrac{1}{3}x^3$, which approaches $\tfrac{1}{3}x^3$ as $h \to 0$, giving the required result.

Exercise 9.2

Area under y=xⁿ with n a positive integer

Consider $\int_0^x t^n\,dt = F(x)$. Use the method of Example 9.1 to show that $F(x) = \tfrac{1}{n+1}x^{n+1}$ for the cases $n = 3, 4,$ and 5. (Use computer algebra for $n \geq 4$.)

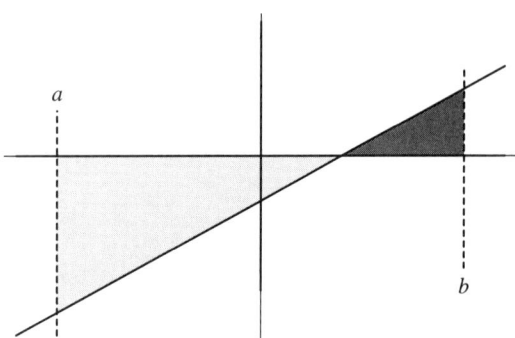

Figure 9.5. Negative area under y=mx+a.

Exercise 9.3

Combining integration intervals

Consider the function $f(x) = \frac{1}{2}x - 1$ on the interval [-2, 2]. Find the value p such that for $x < p$, $f(x) < 0$, and for $x > p$, $f(x) > 0$. Using the function $F(x)$ from Exercise 9.1, verify that result (9.7) holds, and interpret the function $F(x)$ as an area on the graph of $f(x)$.

The results obtained here also apply when the function takes negative values, provided "the area under the curve" is interpreted as a negative quantity when the function takes negative values. This is illustrated in Figure 9.5. Two other important results are:

$$\int_a^b f(x)dx = -\int_b^a f(x)dx \qquad (9.6)$$

$$\int_a^b f(x)dx = \int_a^p f(x)dx + \int_p^b f(x)dx \qquad (9.7)$$

Definite integrals are written as in equations (9.4) to (9.7), with limits (e.g., a and b) shown explicitly. Usually, we think of these values as an interval [a, b] with $a \leq b$ and refer to a as the *lower limit* and b as the *upper limit*, but equation (9.6) tells us that if $b \leq a$, then the enclosed area should be counted as (-1) times the area measured in the direction in which the integration interval increases. Equation (9.7) shows that we may split up the integration interval: the area enclosed over the interval [a, b] equals the sum of the area enclosed over [a, p] plus the area enclosed over [p, b], provided that each area is interpreted as positive or negative in accordance with the sign of $f(x)$ and the rule (9.6).

Antiderivatives and the Fundamental Theorem of Calculus

In the previous section integrals are presented as being related to areas, although the concept of geometrical area (which is always positive) had to be extended to a more general concept of integration area which can take negative as well as positive values. Integration area, as was shown, obeys normal algebraic rules in that it may be added and subtracted. We also obtained some results for integration area in simple cases, and we found functions $F(x)$ having the property that $F'(x) = f(x)$.

The Fundamental Theorem of Calculus (see Chapter 7) tells us that if $F(x)$ is the derivative of $f(x)$ [i.e., $F'(x) = f(x)$], then $\int_a^b f(x)dx = [F(x)]_a^b$. Therefore, to find an integral, rather than going through a complicated process of summing strips of areas as we have just done, we may instead simply try to find $F(x)$ using our knowledge of differentiation of standard functions. Such a function is known as the *antiderivative* of $f(x)$. Since this process has more to do with the form of functions than any particular area over any particular interval [a, b], this relationship is also expressed using a notation called the *indefinite integral*:

$$\int f(x) = F(x) + C \qquad (9.8)$$

Notice the arbitrary constant C. Because the derivative of any constant is zero, it is clear that $\dfrac{d(F(x)+C)}{dx} = \dfrac{dF(x)}{dx}$, and there is therefore no unique antiderivative function $F(x)$ for any given $f(x)$. However, this does not affect the result (9.5) for definite integrals, because the constant C will always cancel.

Exercise 9.4

Nonuniqueness of the antiderivative
Show that result (9.5) remains unchanged if $F(x)$ is replaced by $F(x)+C$.

Example 9.2

Find the antiderivative of x^n.

It is known that $\dfrac{dx^n}{dx} = nx^{n-1}$, so it is expected that the antiderivative of x^n might be related to x^{n+1}. However, $\dfrac{dx^{n+1}}{dx} = (n+1)x^n$, which is the function we want but multiplied by a factor $(n+1)$. Therefore, consider $F(x) = \dfrac{x^{n+1}}{n+1}$. This does have the property that $\dfrac{dF(x)}{dx} = x^n$ and so is an antiderivative. The most general antiderivative is $F(x) = \dfrac{x^{n+1}}{n+1} + C$.

Exercise 9.5

Common antiderivatives
Write down the most general antiderivatives for the functions $\sin(x)$, $\cos(x)$, e^x. Use the software to check your answers.

Exercise 9.6

Airport queuing
Consider the airport queuing situation illustrated in the application section of the software. We use the result from statistics, quoted without derivation or proof, that the distribution of waiting times is described by the exponential distribution. This tells us that the proportion of airplanes whose waiting times lie in the interval $a \le t \le b$ is given by

$$P(a,b) = \int_a^b A e^{-pt} dt.$$

Here, p is a parameter that controls the shape of the distribution curve: the larger p, the less the proportion of airplanes waiting at longer times. For any given p, A is chosen so that the total area under the curve on the interval $[0, \infty]$ is 1: i.e., the probability that an airplane must wait for a time between zero and infinity must be 1. In the application, the integral was approximated by a bar chart. By differentiation, show that $-\dfrac{A}{p} e^{-pt}$ is a suitable antiderivative and hence that

$$P(a,b) = \left[-\dfrac{A}{p} e^{-pt} \right]_a^b = -\dfrac{A}{p}\left(e^{-pb} - e^{-pa} \right).$$

WORKED EXAMPLE 9.1
TWO BRANCHES OF CALCULUS

Here are two examples of the solution of a simple problem. Previously (in Chapter 7, for instance), we met the mathematical relation between acceleration $a(t)$, velocity $v(t)$, and displacement (distance measured positively upward) $s(t)$ for an object moving in a straight line. Here, and in the software, we look at the case of a stone moving up and down under constant gravity. First, when the height $s(t)$ is a known function, we deduce the acceleration $a(t)$ using differential calculus. Second, when $a(t)$ is known, we deduce the height $s(t)$ using integral calculus. We are working in units of feet and seconds.

(i) We use the relationships
$$v(t) = \frac{ds}{dt}$$
$$a(t) = \frac{dv}{dt}.$$

Suppose that from observation
$$s(t) = -16t^2 + 200t.$$
Find the initial position and velocity.

First set $t=0$ to give the initial position $s(0)=0$.

Differentiating gives first
$$v(t) = -32t + 200$$
and then differentiating again gives
$$a(t) = -32.$$
Note that we can now find the velocity and displacement at any time by substituting a value for t. In particular, the initial velocity is 200 ft/sec. Note also that the acceleration is constant.

(ii) We start with the relationship
$$a(t) = \frac{dv}{dt}.$$
Therefore, we may write either
$$v(t) = \int a(t)dt$$
or
$$v(t) = v_0 + \int_0^t a(t)dt.$$
If the upper limit in this integral is taken equal to the lower limit ($t = 0$ in this case), then the integral evaluates to zero and we have $v(t) = v_0$. In general, we may choose any suitable value for the lower limit. Similarly, we may then write either
$$s(t) = \int v(t)dt$$
or
$$s(t) = s_0 + \int_0^t v(t)dt.$$

Find $s(t)$ given that $a(t)$ is constant ($a=-32$ ft/sec/sec), $s(0)=0$ and the initial velocity is 200 ft/sec.

As $a(t)$ is constant, the indefinite integral is known, and we have
$$\int_0^t a\,dt = [at]_0^t = at.$$
Therefore,
$$v(t) = v_0 + \int_0^t a(t)dt = 200 - 32t.$$
Alternatively, use
$$v(t) = \int a\,dt = at + C$$
and since $v(0)=200$ we must have $C=200$, giving
$$v(t) = 200 - 32t$$
which is the same result as before. Finally, since $s(0)=0$ we have
$$s(t) = \int_0^t v(t)dt = \left[200t - 16t^2\right]_0^t$$
and therefore,
$$s(t) = 200t - 16t^2.$$

Numerical Methods

Unfortunately, apart from some simple cases such as functions which are immediately recognizable as derivatives of simple functions, and a number of standard types that are given in textbooks, in general it is not possible to find a function, composed of simple functions, which is an antiderivative. To evaluate such integrals requires the use of a numerical method.

Modern numerical algorithms can be very quick and accurate. Although they rely on theory that is more advanced than can be dealt with in this volume, it is possible to demonstrate some aspects of the theory which affect the efficiency of numerical schemes.

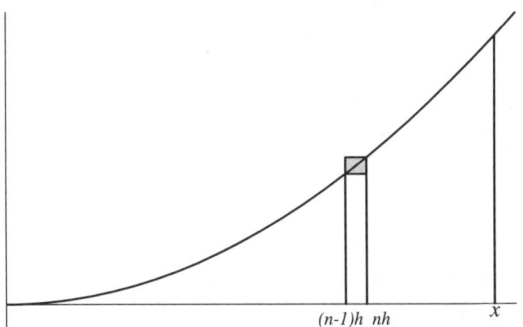

Figure 9.6. Estimating errors.

Consider

$$I = \int_a^b f(x)dx. \qquad (9.9)$$

Numerical methods for finding definite integrals have in effect already been introduced; in Chapter 6, you have seen software demonstrations that use upper and lower sums to estimate integrals, and in this chapter we have already discussed both upper and lower sums, and the trapezoidal method (in Example 9.1). Although in this chapter we have used these methods as analytical tools for finding antiderivatives, each provides a basis for calculation: if a numerical value for h is chosen, then for any given integrand and integration interval, the sums can be evaluated.

The software in the numerical methods concept allows you to compare the performance of these two methods. The key indicator of performance is the way the error changes with h. Suppose we obtain $S(h)$ as our result using the upper or lower sum method. The shaded rectangle in Figure 9.6 shows the difference between the upper and lower sum contribution to an area, and its magnitude therefore gives an upper limit to the magnitude of the error involved. We can estimate this magnitude, because the width of the rectangle is h, and the height will be less than Mh where M is an upper estimate for the magnitude of the slope of the function. Suppose we take

$$M = \max_{a \leq x \leq b} |f'(x)|. \qquad (9.10)$$

This upper estimate holds for all the rectangles in the integration interval. Let E be the total error in $S(h)$, the approximation to the integral I, defined as:

$$E = S(h) - I. \qquad (9.11)$$

Then

$$|E| \leq \frac{|b-a|}{h} Mh^2 = |b-a|Mh \qquad (9.12)$$

because there are $\dfrac{b-a}{h}$ rectangles and each has area at most Mh^2. This is an upper estimate, and the actual error will be different. Nevertheless, equation (9.12) tells us to expect that the error will be proportional to h.

For the trapezoidal method we will not attempt a general result, but simply note that in the specific case used in Example 9.1, the estimate obtained, $T(h)$ say, was

$T(h) = \frac{1}{3}x^3 + \frac{1}{6}h^2 x$, and the exact result was $I = \frac{1}{3}x^3$. Since we took $a=0$ and $b=x$ in this case, we can write

$$|E| = \frac{1}{6}|b-a|h^2. \quad (9.13)$$

This suggests that the error for the trapezoidal method decreases more rapidly than the error for the upper or lower sums method, and that some error bound might be found which is proportional to h^2. This is consistent with the result obtained in the New Situations section at the end of Chapter 8.

Exercise 9.7

Numerical errors

Use the software in the numerical methods concept to compare the accuracy of numerical estimates obtained using upper and lower sums with those obtained using the trapezoidal rule: i.e., compare the accuracy of estimates $S(h)$ with the accuracy of estimates $T(h)$ based on a numerical version of the technique used in Example 9.1. Do not expect that the errors will be found to behave exactly as suggested by theory, because theory only gives upper estimates. You should look for consistency with the predictions above that the error bounds are proportional to h or h^2.

Standard Techniques

The two preceding sections, on Definite Integrals and Area, and Antiderivatives and the Fundamental Theorem of Calculus, have revised materials from earlier chapters, 6 and 7 in particular. They can be summarized very briefly by saying that to find an area under a curve $y = f(x)$ (suitably interpreting area), it is only necessary to find an antiderivative $F(x)$ and then apply equation (9.5):

$$\int_a^b f(x)dx = [F(x)]_a^b \quad (9.5)$$

This "only" leaves the problem of finding an antiderivative $F(x)$. Finding antiderivatives is best thought of as an art analogous to solving a crossword puzzle, rather than as a science. Computer algebra systems do make use of an elaborate algorithm for finding an antiderivative (if one exists), but this is not a good basis for pencil and paper working.

We need to consider exactly what we mean by "finding an antiderivative". Without using rigorous mathematics, it is clear that any reasonably smooth function whose graph we can plot will have a well-defined function that describes "the area under the curve." So in that sense, an antiderivative must exist. The point is, can this function be expressed in terms of previously known simple mathematical functions? Now, we must define what we mean by "simple functions." It is usually considered that these are polynomials, rational functions, and the elementary functions $\sin(x)$, $\cos(x)$, e^x and their inverses. They can be composed into an infinite number of combinations, including all the other trigonometric functions. Notice that $\ln(x)$ is the inverse of e^x, so this is included. Here are some examples of the way these functions can be composed:

$$e^x \cos(e^x), \quad \frac{\ln(\alpha x) + \cos(\beta x^2)}{1+x}, \quad e^{-x^2}.$$

The first of these, you may see quickly is the derivative of $\sin(e^x)$, while the second and third have no obvious indefinite integral. In fact, no "simple" antiderivatives exist for these functions, and we say that

STANDARD INTEGRALS

$$\int x^n dx = \frac{x^{n+1}}{n+1} + C, \ n \neq -1 \quad \text{since} \quad \frac{d}{dx}\left(\frac{x^{n+1}}{n+1}\right) = x^n$$

$$\int \frac{1}{x} dx = \log(x) + C \quad \text{since} \quad \frac{d}{dx}(\log(x)) = \frac{1}{x}$$

$$\int \cos(mx) dx = \frac{1}{m}\sin(mx) + C, \quad \text{since} \quad \frac{d}{dx}\left(\frac{1}{m}\sin(mx)\right) = \cos(mx)$$

$$\int \sin(mx) dx = -\frac{1}{m}\cos(mx) + C, \quad \text{since} \quad \frac{d}{dx}\left(-\frac{1}{m}\cos(mx)\right) = \sin(mx)$$

$$\int e^x dx = e^x + C \quad \text{since} \quad \frac{d}{dx}(e^x) = e^x$$

$$\int \log(x) dx = x\log(x) - x + C \quad \text{since} \quad \frac{d}{dx}(x\log(x) - x) = \log(x)$$

their antiderivatives cannot be expressed "in closed form." The function $\frac{e^{-x^2}}{\sqrt{\pi}}$ describes the "bell curve" and is so useful in Statistics that it is used to describe a "new" function:

$$\text{erf}(x) = \frac{1}{\sqrt{\pi}} \int_{-x}^{x} e^{-t^2} dt.$$

The search for antiderivatives begins simply by looking at a list of standard cases. This Laboratory Manual is not intended as a reference work, and you should consult standard textbooks for tables of Standard Integrals. The table, however, contains a list of some of the most commonly met cases.

Rules of Integration

Chapter 3 (Rates of Change and Differentiation) showed that there are certain simple rules that help in manipulating functions when differentiating them. For example,

$$\frac{d(kf)}{dx} = k\frac{df}{dx}$$

where f is a function of x and k is a constant.

There are equivalent rules for integration, which help with "the art of finding antiderivatives." These rules are explained in the software and summarized below for reference.

Constant Multiple Rule

The constant multiple rule for integrals states that if $f(x)$ is an integrable function (i.e., $\int f(x)dx$ exists) and k is a constant, then

$$\int kf(x)dx = k\int f(x)dx.$$

It follows also that for a definite integral over any interval $[a, b]$,

$$\int_a^b kf(x)dx = k\int_a^b f(x)dx.$$

This rule is obtained by considering an integral as the limit of the sum of areas of strips, as in the section Definite Integrals and Area earlier in this chapter. The constant factor k multiplies the area of each strip by k. Hence the total area, and the limit of the area, are also multiplied by k.

Sum Rule

The sum rule for integrals states that if $f(x)$ and $g(x)$ are integrable functions (i.e., $\int f(x)dx$ and $\int g(x)dx$ exist), then

$$\int (f(x) \pm g(x))dx = \int f(x)dx \pm \int g(x)dx.$$

It follows also that for a definite integral over any interval $[a, b]$,

$$\int_a^b (f(x) \pm g(x))dx = \int_a^b f(x)dx \pm \int_a^b g(x)dx.$$

Again, this rule is obtained by considering an integral as the limit of the sum of areas of strips. The height of each strip in a graph of $f(x) \pm g(x)$ is the sum (or difference) of $f(x)$ and $g(x)$. Hence the area under the curve is the same as the sum (or difference) of the areas under the curves $f(x)$ and $g(x)$ separately.

Integration by Parts

This rule states that if $f(x)$ and $g(x)$ are differentiable and integrable functions (i.e., $f'(x)$, $g'(x)$, $\int f(x)dx$ and $\int g(x)dx$ exist), then

$$\int f(x)g'(x)dx = f(x)g(x) - \int g(x)f'(x)dx.$$

It follows also that for a definite integral over any interval $[a, b]$,

$$\int_a^b f(x)g'(x)dx = [f(x)g(x)]_a^b - \int_a^b g(x)f'(x)dx.$$

This rule is derived from the product rule for differentiation. Let

$$u(x) = f(x)g(x).$$

Then

$$u'(x) = f'(x)g(x) + f(x)g'(x).$$

But

$$\int_a^b u'(x)dx = [u(x)]_a^b$$

which gives the result on substituting for $u(x)$ and $u'(x)$.

An alternative form uses differential notation as follows. In this form, it is allowed to write quantities like df, which stand for Δf, with the implied assumption that eventually we will take limits like $\frac{\Delta f}{\Delta x} \to \frac{df}{dx}$. Using this notion, we may write

$$f'(x)dx = \frac{df}{dx}dx = df$$

and

$$g'(x)dx = \frac{dg}{dx}dx = dg.$$

The indefinite form of the rule for integration by parts may then be written:

$$\int f(x)dg = f(x)g(x) - \int g(x)df.$$

Substitution

This technique is based on the chain rule for differentiation. If u is a differentiable function of x, then

$$\int f(u)\frac{du}{dx}dx = \int f(u)du.$$

To derive the rule, suppose that F is an antiderivative of f, so that

$$\frac{dF(u)}{du} = f(u),$$

or equivalently

$$\int f(u)du = F(u)+C.$$

From the chain rule,

$$\frac{dF(u)}{dx} = F'(u)\frac{du}{dx} = f(u)\frac{du}{dx}$$

so we also have

$$\int f(u)\frac{du}{dx}dx = F(u)+C$$

and the result follows.

The form for definite integrals is

$$\int_a^b f(u)\frac{du}{dx}dx = \int_{u(a)}^{u(b)} f(u)du.$$

Notice that the limits for u are the values that correspond to the limits for x.

The point of substitution is that often it can transform an otherwise impossible looking integral into something more familiar. There is no fixed rule that can be given to suggest the best choice, but the art of integration improves with practice and experience.

Using the differential notation mentioned in the preceding section Integration by Parts, we see that the result is immediate.

WORKED EXAMPLE 9.2
RULES OF INTEGRATION

(i) Find $I = \int_1^2 (2x^2 + 3x^3) dx$.

First, use the sum rule to give

$$I = \int_a^b 2x^2 dx + \int_a^b 3x^3 dx$$

and then use the constant multiple rule for each term to give

$$I = 2\int_a^b x^2 dx + 3\int_a^b x^3 dx.$$

We know the antiderivative of x^n is $\frac{x^{n+1}}{n+1}$, so

$$I = 2\left[\frac{x^3}{3}\right]_1^2 + 3\left[\frac{x^4}{4}\right]_1^2 = \left[\frac{2}{3}x^3 + \frac{3}{4}x^4\right]_1^2$$

which evaluates to $15\frac{11}{12}$.

(ii) Find $I = \int_0^1 \frac{x}{\sqrt{5x^2 + 1}} dx$.

Substitute
$$u = 5x^2 + 1.$$

We have
$$\frac{du}{dx} = 10x,$$

and therefore the substitution rule gives

$$I = \frac{1}{10}\int_0^1 u^{-\frac{1}{2}} \frac{du}{dx} dx = \frac{1}{5}\int_{u(0)}^{u(1)} u^{\frac{1}{2}} du.$$

$$\therefore I = \frac{1}{5}\left[u^{\frac{1}{2}}\right]_{u(0)}^{u(1)} = \frac{1}{5}\left[\sqrt{5x^2 + 1}\right]_0^1.$$

This evaluates to $I = \frac{1}{5}(\sqrt{6} - 1)$.

(iii) Find $I = \int_2^3 \frac{dx}{5x - 3}$.

Substitute $u = 5x - 3$. We have $\frac{du}{dx} = 5$, and therefore the substitution rule gives

$$I = \frac{1}{5}\int_2^3 \frac{1}{u} \frac{du}{dx} dx = \frac{1}{5}\int_{u(2)}^{u(3)} \frac{du}{u}.$$

$$\therefore I = \frac{1}{5}\left[\ln(u)\right]_{u(2)}^{u(3)} = \frac{1}{5}\left[\ln(5x - 3)\right]_2^3.$$

This evaluates to $I = \frac{1}{5}\ln\left(\frac{12}{7}\right)$.

(iv) Find $I = \int_0^a \cos^2(x) dx$.

Use the trigonometric identity
$$\cos^2(x) = \frac{1}{2}(1 + \cos(2x)).$$

Now we can apply the sum rule, constant multiple rule and substitution rule to give

$$I = \left[\frac{1}{2}\left(x + \frac{1}{2}\sin(2x)\right)\right]_0^a = \frac{1}{2}\left(a + \frac{1}{2}\sin(2a)\right).$$

(v) Find $\int_a^b \frac{dx}{1 - x^2}$, where $-1 < a < b < 1$.

Integrals of this type, where the denominator can be factorized, may often be solved by using partial fractions (see standard textbooks for details of the technique). In this example, we note

$$\frac{1}{1 - x^2} = \frac{1}{2}\left[\frac{1}{1 + x} + \frac{1}{1 - x}\right].$$

$$\therefore \int_a^b \frac{dx}{1 - x^2} = \frac{1}{2}\left[\int_a^b \frac{dx}{1 + x} + \int_a^b \frac{dx}{1 - x}\right]$$

$$= \frac{1}{2}\left[\ln(1 + x) - \ln(1 - x)\right]_a^b$$

$$= \frac{1}{2}\ln\left(\frac{(1 + b)(1 - a)}{(1 - b)(1 + a)}\right).$$

(vi) Find $I = \int_0^\alpha \tan(\theta) d\theta$.

Since $\tan(\theta) = \frac{\sin(\theta)}{\cos(\theta)}$, try the substitution $u = \cos(\theta)$. Now $\frac{du}{d\theta} = -\sin(\theta)$, therefore,

$$I = \int_0^\alpha -\frac{1}{u}\frac{du}{d\theta} d\theta.$$

Applying the substitution rule gives

$$I = \int_1^{\cos(\alpha)} -\frac{1}{u} du = \int_{\cos(\alpha)}^1 \frac{1}{u} du.$$

This is now a standard form and we have

$$I = \left[\ln(u)\right]_{\cos(\alpha)}^1 = -\ln(\cos(\alpha)).$$

(vii) Find $I = \int_0^2 \ln(x)dx$.

The antiderivative for $\ln(x)$ is not immediately obvious (unless you know it in advance), but the derivative of $\ln(x)$ is well known. This is an place to try integration by parts. The rule is given by

$$\int_a^b f(x)g'(x)dx = [f(x)g(x)]_a^b - \int_a^b g(x)f'(x)dx.$$

If we set $f(x) = \ln(x)$, then there is no obvious choice for $g'(x)$. Let us try $g'(x) = 1$, so that $g(x)=x$ is a valid choice for $g(x)$. Therefore,

$$I = [x\ln(x)]_0^2 - \int_0^2 x\frac{1}{x}dx = [x\ln(x) - x]_0^2.$$

Using $x\ln(x) \to 0$ as $x \to 0$ gives

$$I = 2\ln(2) - 2.$$

(viii) Find $I = \int_0^1 (1-x^2)^{\frac{1}{2}} dx$.

When a term like $(1-x^2)$ is present, a trigonometric substitution may help, using, for example, $\cos^2(u) = 1 - \sin^2(u)$. We try

$$x = \sin(u).$$

Then $1 = \cos(u)\dfrac{du}{dx}$, and therefore

$$I = \int_0^1 (1-\sin(u)^2)^{\frac{1}{2}} \cos(u) \frac{du}{dx} dx$$

$$\therefore I = \int_a^b \cos^2(u)du$$

where $u=a$ when $x=0$ and $u=b$ when $x=1$, so that $a=0$ and $b=\frac{\pi}{2}$.

Alternatively, using the differential notation mentioned earlier under Integration by Parts, we find that the above appears as follows. Let

$$x = \sin(u).$$

$$\therefore dx = \cos(u)du$$

$$I = \int_0^1 (1-\sin^2(x))^{\frac{1}{2}} \cos(u)du = \int_0^1 \cos^2(u)du.$$

The result from part (iv) above for $\int_0^a \cos^2(x)dx$ gives

$$I = \left[\tfrac{1}{2}\left(u + \tfrac{1}{2}\sin(2u)\right)\right]_0^{\frac{\pi}{2}} = \tfrac{\pi}{4}.$$

(ix) Find $I = \int_{-3}^1 \dfrac{x+3}{\sqrt{7-6x-x^2}} dx$.

With a denominator of this form it is often useful to complete the square, that is, write

$$7 - 6x - x^2 = 16 - (3+x)^2.$$

Now make the substitution $u = x+3$. This gives

$$\sqrt{7-6x-x^2} = \sqrt{16-(3+x)^2} = \sqrt{16-u^2}.$$

Then

$$I = \int_0^4 \frac{u\,du}{\sqrt{16-u^2}}.$$

A further substitution may then be made:

$$v(u) = \sqrt{16-u^2}.$$

Note that $v(0)=4$ and $v(4)=0$. Using differential notation we have

$$dv = -\frac{u}{\sqrt{16-u^2}}du$$

and therefore

$$I = -\int_4^0 dv = 4.$$

(x) Show $\int \sec(\theta)d\theta = \ln(\sec(\theta)+\tan(\theta))$.

This is a case where an inspired guess is needed if the indefinite integral were not known. To verify the result we need:

$$\frac{d\tan(\theta)}{d\theta} = \sec^2(\theta)$$

$$\frac{d\sec(\theta)}{d\theta} = \sec(\theta)\tan(\theta).$$

Starting from the right-hand side,

$$\frac{d[\ln(\sec(\theta)+\tan(\theta))]}{d\theta}$$

$$=\frac{\sec(\theta)\tan(\theta)+\sec^2(\theta)}{\sec(\theta)+\tan(\theta)}$$

$$=\sec(\theta).$$

The right-hand side is therefore an antiderivative for the integrand, and we have the result.

(xi) Find the arc length of the sector of the circle

$$y=\sqrt{1-x^2}$$

from $x=0$ to $x=1$.

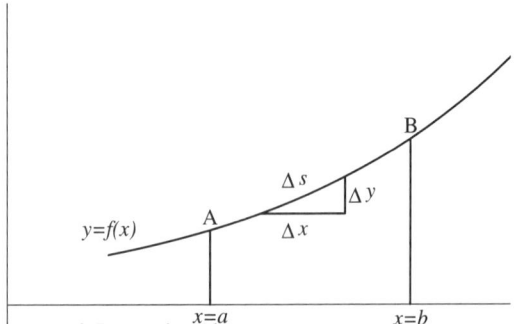

Figure 9.7. Arc length.

The arc length along any curve is found by integration (see Figure 9.7). On the curve $y=f(x)$, let Δs be the length of a small line segment from the point (x, y) to $(x+\Delta x, y+\Delta y)$. Therefore,

$$\Delta s^2 = \Delta x^2 + \Delta y^2$$

$$= \left(1+\left(\frac{\Delta x}{\Delta y}\right)^2\right)\Delta x^2$$

$$\therefore \Delta s = \left(1+\left(\frac{\Delta x}{\Delta y}\right)^2\right)^{\frac{1}{2}}\Delta x$$

We also know that in the limit as $\Delta x \to 0$, $\frac{\Delta y}{\Delta x} \to \frac{dy}{dx}$. In the indefinite form,

$$\text{arc length} = \int ds.$$

The arc length from $x=a$ to $x=b$ is therefore given by

$$L(a,b)=\int_a^b \sqrt{1+\left(\frac{dy}{dx}\right)^2}\,dx.$$

For the circle sector in this problem,

$$\frac{dy}{dx}=-x(1-x^2)^{-\frac{1}{2}}.$$

$$\therefore L(0,1)=\int_0^1 \left(1-x^2\right)^{-\frac{1}{2}}dx.$$

This integral is similar to the integral solved in part (viii) above and may be found by the same substitution. Let

$$x=\sin(\theta)$$

which gives

$$dx=\cos(\theta)d\theta$$

and corresponding to the interval [0, 1] for x, we have [0, $\frac{\pi}{2}$] as the interval for q. Using this and the identity

$$\cos^2(\theta)+\sin^2(\theta)=1$$

gives

$$L(0,1)=\int_0^{\pi/2} d\theta = \frac{\pi}{2}.$$

It is worth noting that the circle of which $y=\sqrt{1-x^2}$ is a sector has radius 1 and circumference 2π. The sector of the curve from $x=0$ to $x=1$ is a quarter circle, and therefore our result is as expected.

PROBLEM 9.1
DEFINITE INTEGRATION

Name: _____

Date: _____

Section: _____

Problem 9.1a

Recall the airport queuing application and the results of Exercise 9.6. This tells us that the proportion of airplanes whose waiting times lie in the interval $a \le t \le b$ *is given by*

$$P(a,b) = \int_a^b A e^{-pt} dt$$

and that the definite integral can be found to be

$$P(a,b) = \left[-\frac{A}{p} e^{-pt} \right]_a^b = -\frac{A}{p}\left(e^{-pb} - e^{-pa}\right).$$

(i) Use the software in the airport queuing application. Collect some data, and choose values for A and p. Sketch the distribution Ae^{-pt} and record your values here.

A = ...   p = ...

(ii) Using these values, what proportion of planes will have to wait:

more than 10 minutes?   ....

less than 10 minutes?   ....

(iii) Add your two answers just obtained. Record their sum: ....

Expressed in proportions, the sum ought to be 1, because all planes have either to wait more than 10 minutes or less than 10 minutes! This means that $P(0, \infty) = 1$ should always hold. Use the results quoted on this page to write down a relation between A and p that must be satisfied for this condition to hold:

$$P(0, \infty) = \int_0^\infty A e^{-pt} dt = 1 \quad \Rightarrow \quad \text{............................}$$

PROBLEM 9.1
DEFINITE INTEGRATION
(Continued)

Problem 9.1b
Sketch each integrand given below. Use pencil and paper to find mathematical formulas for the definite integrals, and evaluate them for the intervals given.

Then use the software in the Numerical Methods section to evaluate these integrals using the trapezoidal rule, with the values of h suggested. Record your results, and calculate the error as the difference between the estimated value and the exact value of the integral.

(i) $I = \int_a^b (1+2x)dx$ = ... $f(x)$

For $[a, b] = [0, 1]$, I = ...

	h = 0.2	0.1	0.05	0.025
Estimate for I =				
Error =				

(ii) $I = \int_a^b (12-3x^2)dx$ = ... $f(x)$

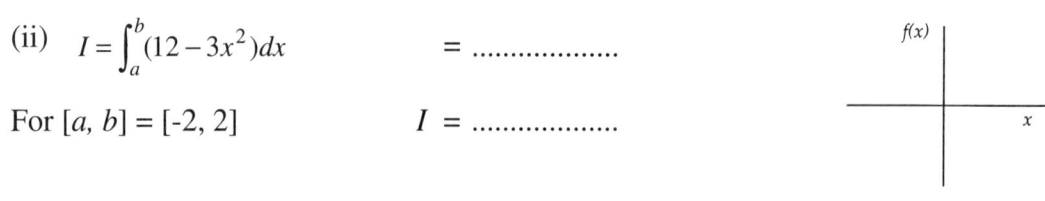

For $[a, b] = [-2, 2]$ I = ...

	h = 0.2	0.1	0.05	0.025
Estimate for I =				
Error =				

(iii) $I = \int_a^b (2x^3 - 4x^2 - 10x + 12)dx$ = ... $f(x)$

For $[a, b] = [-2, 3]$ I = ...

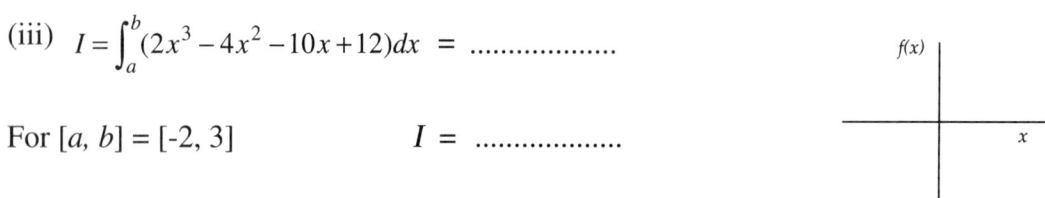

	h = 0.2	0.1	0.05	0.025
Estimate for I =				
Error =				

PROBLEM 9.2
RULES OF INTEGRATION

Name: _____

Date: _____

Section: _____

Apply standard techniques and rules of integration to solve the following problems.

(i) Use a trigonometric substitution or partial fractions to evaluate $\int_{-2}^{2} \frac{dx}{9-x^2}$.

(ii) Use integration by parts to find $\int_{0}^{\pi} x^3 \sin(x)\,dx$.

(iii) Use integration by parts to find $\int_{0}^{1} a^x\,dx$. (*Hint:* if $u = a^x$ first find $\frac{du}{dx}$.)

(iv) Use the substitution $t=\tan(x)$ to show that $\int_{0}^{\frac{\pi}{4}} \frac{dx}{1+\sin(2x)} = \frac{1}{2}$ and find $\int_{0}^{\frac{\pi}{4}} \frac{dx}{(1+\sin(2x))^2}$.

PROBLEM 9.2
RULES OF INTEGRATION
(Continued)

(v) Using the substitution $x = 1 - \dfrac{1}{u^4}$, show that $\displaystyle\int_1^{(2)^{\frac{1}{4}}} \dfrac{du}{u(2u^4-1)^{\frac{1}{4}}} = \dfrac{\pi}{24}$.

(vi) Show that $\displaystyle\int_a^b \dfrac{x\,dx}{\sqrt{(x-a)(b-x)}} = \dfrac{1}{2}\pi(a+b)$.

(*Hint:* use the substitution $x = a\cos^2(\theta) + b\sin^2(\theta)$.)

(vii) Let $I_n = \displaystyle\int_0^{\pi} \sin^n(\theta)d\theta$ where n is a positive integer. Use integration by parts to show that $I_n = \dfrac{n-1}{n} I_{n-2}$. Obtain a similar formula for $J_n = \displaystyle\int_0^{\pi} \cos^n(\theta)d\theta$ and evaluate I_4 and J_4.

(*Hint:* see reduction formulas in the New Situations section.)

(viii) Let $u_n = \displaystyle\int_0^{\frac{\pi}{2}} \dfrac{\sin(2n\theta)d\theta}{\sin(\theta)}$ where n is a positive integer. Show that $u_n - u_{n-1} = \dfrac{2(-1)^{n-1}}{2n-1}$.

Hence prove that $u_n = 2\left\{1 - \dfrac{1}{3} + \dfrac{1}{5} - \ldots + \dfrac{(-1)^{n-1}}{2n-1}\right\}$.

NEW SITUATIONS

1. Reduction formulas

If $I_n = \int_a^b \sec^n(\theta)d\theta$, show that

$$I_n = \frac{1}{n-1}\left[\sec^{n-2}(\theta)\tan(\theta)\right]_a^b - \frac{n-2}{n-1}I_{n-2}$$

where n is a positive integer.

This is an example of a class of formulas known as *reduction formulas*. In most instances, integration by parts will give the result, although the choice of $f(x)$ and $g(x)$ remains an art. In this case the choice is suggested by calling on the following standard results:

$$\sec^2(\theta) - \tan^2(\theta) = 1$$

$$\frac{d\tan(\theta)}{d\theta} = \sec^2(\theta)$$

$$\frac{d\sec(\theta)}{d\theta} = \sec(\theta)\tan(\theta).$$

We have:

$$I_n = \int_a^b \sec^n(\theta)d\theta$$

$$= \int_a^b \sec^{n-2}(\theta)\sec^2(\theta)d\theta$$

$$= \int_a^b \sec^{n-2}(\theta)\frac{d\tan(\theta)}{d\theta}d\theta$$

$$= \left[\sec^{n-2}(\theta)\tan(\theta)\right]_a^b$$
$$\qquad - \int_a^b \frac{d\sec^{n-2}(\theta)}{d\theta}\tan(\theta)d\theta$$

$$= \left[\sec^{n-2}(\theta)\tan(\theta)\right]_a^b$$
$$\qquad -(n-2)\int_a^b \sec^{n-3}(\theta)(\sec(\theta)\tan(\theta))\tan(\theta)d\theta$$

$$= \left[\sec^{n-2}(\theta)\tan(\theta)\right]_a^b$$
$$\qquad -(n-2)\int_a^b \sec^{n-2}(\theta)\left(1+\sec^2(\theta)\right)d\theta$$

$$= \left[\sec^{n-2}(\theta)\tan(\theta)\right]_a^b - (n-2)I_n - (n-2)I_{n-2}$$

$$\therefore I_n + (n-2)I_n =$$
$$\qquad \left[\sec^{n-2}(\theta)\tan(\theta)\right]_a^b - (n-2)I_{n-2}$$

$$\therefore I_n = \frac{1}{n-1}\left[\sec^{n-2}(\theta)\tan(\theta)\right]_a^b - \frac{n-2}{n-1}I_{n-2}$$

as required.

For even n, repeated application of this formula eventually produces an expression in which the only term involving an unknown integral is multiplied by zero and hence vanishes. For odd n, the problem is eventually reduced to finding the integral for the case $n=1$. In Worked Examples 9.2 we obtained the indefinite form of the integral for $n=1$, giving

$$I_1 = \int_a^b \sec(\theta)d\theta = \left[\ln(\sec(\theta)+\tan(\theta))\right]_a^b.$$

For example,

$$I_4 = \frac{1}{3}\left[\sec^2(\theta)\tan(\theta)\right]_a^b - \frac{2}{3}I_2$$

$$I_2 = \left[\tan(\theta)\right]_a^b - \frac{0}{1}I_0$$

gives I_4, and

$$I_3 = \frac{1}{2}\left[\sec(\theta)\tan(\theta)\right]_a^b - \frac{1}{2}I_1$$

gives I_3 in terms of I_1 which we know.

2. Arc length for a parabola

To find the arc length of the parabola
$$y = kx^2$$
from $x=a$ to $x=b$, start with the general result found in example (xi) of Worked Examples 9.2:

$$L(a,b) = \int_a^b \sqrt{1+\left(\frac{dy}{dx}\right)^2}\,dx.$$

For the parabola in this problem,

$$\frac{dy}{dx} = 2kx.$$

$$\therefore L(a,b) = \int_a^b \left(1 + 4k^2 x^2\right)^{\frac{1}{2}} dx.$$

Now make the substitution

$$4kx = \tan(\theta)$$

which gives

$$4k\, dx = \sec^2(\theta)\, d\theta.$$

Using this and the identity

$$\sec^2(\theta) - \tan^2(\theta) = 1$$

gives

$$L(a,b) = \frac{1}{4k} \int_\alpha^\beta \sec^3(\theta)\, d\theta.$$

Here, $[\alpha, \beta]$ is the interval for θ that corresponds to the interval $[a, b]$ for x, that is, $4ka = \tan(\alpha)$, $4kb = \tan(\beta)$. This integral is $\frac{1}{4k} I_3$ where I_n is the integral found using the reduction formula method described in the previous section.

3. Arc length for parametric curves

When a curve is given parametrically as, for example, $(x(t), y(t))$, then the arc length may be found by a simple adaptation of the method given before.

Let Δs be the length of a small line segment from the point (x, y) to $(x+\Delta x, y+\Delta y)$. This is as before, but now we also have

$$\Delta x = \frac{\Delta x}{\Delta t} \Delta t, \quad \Delta y = \frac{\Delta y}{\Delta t} \Delta t.$$

Then

$$\Delta s^2 = \Delta x^2 + \Delta y^2 = \left\{ \left(\frac{\Delta x}{\Delta t}\right)^2 + \left(\frac{\Delta y}{\Delta t}\right)^2 \right\} \Delta t^2.$$

Taking the limit as $\Delta t \to 0$,

$$L(a,b) = \int_a^b \left\{ \left(\frac{dx}{dt}\right)^2 + \left(\frac{dy}{dt}\right)^2 \right\}^{\frac{1}{2}} dt$$

is the arc length along the portion of the curve from $t=a$ to $t=b$.

Chapter 10 Rectilinear Motion

Prerequisites

Before you study this material, you should be familiar with:
  (1) Differentiation (Module 3).
  (2) Integration (Modules 7 and 9).

Objectives

This chapter illustrates how the Differential and Integral Calculus can be used to investigate motion in straight lines; this is usually called *rectilinear motion*. We have already seen that the Differential Calculus gives us a way to find the velocity and acceleration of a particle given its position as a function of time. Alternatively, given its acceleration at any time, we can use the Integral Calculus to determine its velocity and position. Although the integral calculus introduces arbitrary constants, we will see how specifying the velocity and position of a particle at particular times enables us to determine the position uniquely.

Connections

This module shows how the calculus is used to provide a theoretical basis for the mathematical modeling of moving objects. Although many simplifying assumptions are made, the basic principles carry through into more advanced applications. For example, the Apollo Space Program used the same basic theory to construct navigational software, although the engineers had to use a complex mathematical model of the moon's gravity to get the lunar landing trajectory right. (Apollo 11 landed 20 kilometers from its intended landing site, but the model was improved for later missions.)

The two videos associated with this chapter investigate the motion of a dragster and a parachute. The dragster is modeled very simply using constant acceleration and deceleration. Effects like air resistance and the fact that engines do not deliver constant acceleration are ignored, for example. The parachute application starts with free fall under gravity with no air resistance; then after the parachute opens, air resistance is assumed to be proportional to the velocity of the sky diver.

The animations are designed to reinforce the ideas that derivatives measure rates of change and integrals are related to areas under curves.

General Results

The word rectilinear just means "straight line", and although there are very few situations in which straight line motion is adequate to describe real-world events, we have to be able to cope with this simple case before we can go on to handle more complex motion.

If a particle is moving in a straight line, then its positions can be specified by a function $x(t)$ which gives its position in terms of the distance from a fixed point. The velocity $v(t)$ is given by the derivative of $x(t)$ with respect to time:

$$v(t) = \frac{dx}{dt} \quad (10.1)$$

and its acceleration, $a(t)$, is given by

$$a(t) = \frac{dv}{dt} = \frac{d^2x}{dt^2} \quad (10.2)$$

Alternatively, suppose that the acceleration at any time t is given by $a(t)$, and its position and velocity are known at time $t = t_0$ given by $x(t_0) = x_0$, and $v(t_0) = v_0$; then we may integrate (10.2) to give:

$$\int_{t_0}^{t} a(t)dt = \int_{t_0}^{t} \frac{dv}{dt} dt = \int_{t_0}^{t} dv.$$

Therefore,

$$[v(t)]_{t_0}^{t} = v(t) - v_0 = \int_{t_0}^{t} a(t)dt$$

or

$$v(t) = v_0 + \int_{t_0}^{t} a(t)dt.$$

If $a(t)=a$, where a is constant, then we may integrate this expression to get

$$x(t) = x_0 + (t - t_0)a \quad (10.3)$$

Similarly, (10.1) may be integrated to give

$$\int_{t_0}^{t} v(t)dt = \int_{t_0}^{t} \frac{dx}{dt} dt$$

and so

$$x(t) = x_0 + \int_{t_0}^{t} v(t)dt.$$

Again, if $a(t)=a$, a constant, then we can integrate equation (10.3) to get

$$x(t) = x_0 + \left[v_0 t + \tfrac{1}{2} a(t - t_0)^2\right]_{t_0}^{t}$$

$$= x_0 + v_0(t - t_0) + \tfrac{1}{2} a(t - t_0)^2.$$

Notice that in the above expression the time always appears in the results as the factor $(t - t_0)$. We can interpret this as "time since the initial conditions at $t = t_0$", and often without loss of generality, we can define $t_0 = 0$ and simply use t in place of $(t - t_0)$.

Example 10.1

A stone is thrown vertically upward so that its height in feet after t seconds is given by

$$h(t) = -16t^2 + 200t.$$

What are the velocity and acceleration of the stone after 10 seconds?

The velocity of the stone is given by

$$v(t) = \frac{dh}{dt} = -32t + 200$$

so that after 10 seconds the velocity is -120 feet per second. That is, the stone is falling at a speed of 120 feet per second. Similarly, the acceleration of the stone is given by

$$a(t) = \frac{dv}{dt} = \frac{d^2h}{dt^2} = -32.$$

Hence we have constant acceleration.

Example 10.2

A ball is dropped from the edge of a cliff and falls under gravity to a beach 50 meters below.

Determine the velocity when the ball strikes the beach.

If we measure distance vertically down from the point where the ball is dropped, then as the ball falls under gravity its acceleration will be $a(t)$=9.81 m/s². Integrating this expression gives

$$v(t) = 9.81t + v_0,$$

where v_0 is the initial velocity of the ball. The ball is released with v_0=0, and so $v(t) = 9.81t$. Integrating this expression gives

$$x(t) = 4.905t^2 + x_0$$

and if we set $x(0)$=0, then $x_0 = 0$, giving

$$x(t) = 4.905t^2.$$

From this expression we can work out the time, t_1, it takes for the ball to reach the beach since

$$x(t_1) = 4.905t_1^2 = 50$$

which gives

$$t_1^2 = \frac{50}{4.905} \approx 10.19$$

or

$$t_1 \approx \sqrt{10.19} \approx 3.19 \text{ seconds}.$$

Substituting this expression into $v(t) = 9.81t$, we obtain $v(t_1) = 31.29$ m/s. Notice that since we are measuring distance vertically down from the point at which the ball is released, $v(t)$ is positive.

Exercise 10.1

The position of a particle is given by

$$x(t) = 1 + 2\sin(t).$$

Determine the velocity and acceleration of the particle when t=2. When is the particle at rest, and what is its acceleration at such points?

Exercise 10.2

A particle is attracted to the origin so that its acceleration at any time t is $\sin(t)$. Determine the velocity and position of the particle as a function of t if the particle is initially 2 centimeters from the origin and passes through the origin after 2 seconds.

Exercise 10.3

We have seen that integrating the acceleration, $a(t) = \dfrac{dv}{dt}$, we can obtain the velocity, v, as a function of time. Show that if a is constant, then the velocity of a particle which is initially at rest at the origin satisfies $v = \sqrt{2ax}$.

(*Hint*: Show that $\dfrac{dv}{dt} = \dfrac{dv}{dx}\dfrac{dx}{dt} = v\dfrac{dv}{dx}$.)

WORKED EXAMPLE 10.1
DRAGSTER RACING

In this example, we will derive equations for the position and velocity of a dragster given its acceleration and braking. We will assume the acceleration is equal to a constant, *a,* in this phase. Then the driver brakes, and we will assume the braking deceleration is *b,* also constant.

In the software you can alter the acceleration and deceleration by sliding the bars on the graph up or down. The idea is to try and control the dragster so that it covers the quarter mile course in the shortest time but that you live to tell the tale. If you brake too late or not hard enough the car will go off the end of the track, and the results can be very dramatic. When you have completed a run, you can plot the acceleration, velocity, and displacement as shown in Figure 10.1.

First, let us consider the acceleration phase, that is, up to any time before the brakes are applied. Let *a* be the constnat acceleration of the dragster and $v=v(t)$ be its speed at time *t*. We shall assume that the dragster has a standing start so that the initial speed, $v(0)$, is zero.

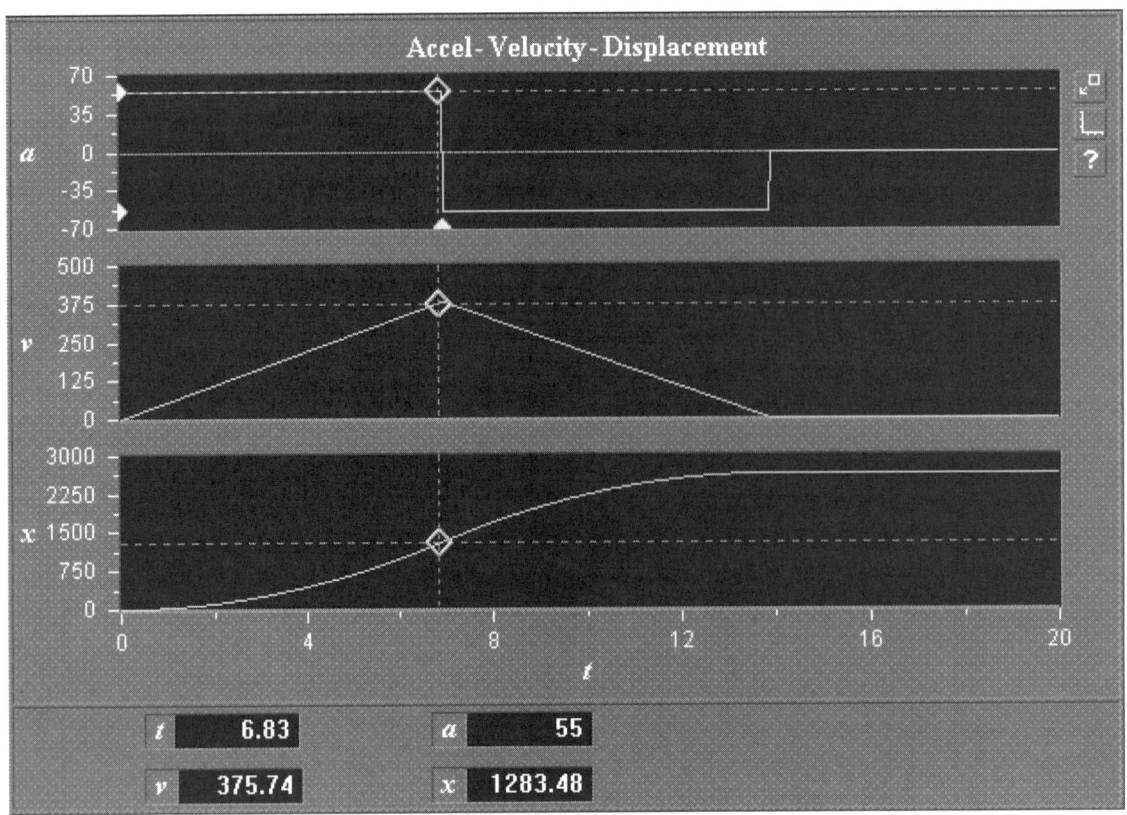

Figure 10.1. *Graphical display of a dragster race.*

Rectilinear motion

Integrating the acceleration $a = \dfrac{dv}{dt}$ with respect to t over the range $t=0$ to t gives

$$v(t) - v(0) = \int_{t=0}^{t} a\, dt = [at]_{0}^{t} = at$$

and so $v(t) = v(0) + at$.

Now let $x(t)$ be the position of the car at time t. We know $v(t) = \dfrac{dx}{dt}$, and therefore we integrate again to give

$$\int_{0}^{t} \frac{dx}{dt}\, dt = \int_{0}^{t} v(t)\, dt = \int_{0}^{t} (v(0) + at)\, dt.$$

Therefore,

$$[x(t)]_{0}^{t} = \left[v(0)t + \tfrac{1}{2}at^2\right]_{0}^{t}$$

hence

$$x(t) - x(0) = v(0)t + \tfrac{1}{2}at^2$$

or

$$x(t) = x(0) + v(0)t + \tfrac{1}{2}at^2.$$

The last result is a general formula for the distance traveled by any object traveling in a straight line under constant acceleration, provided its initial position and speed are known.

In our case we start the stopwatch as the car starts on the starting grid, so $x(0)=0$ and $v(0)=0$, giving $x(t) = \tfrac{1}{2}at^2$.

The second stage of the race is the deceleration phase, from the time that braking starts. Suppose this time is $t=T$. Let $v=v(T)$ be the speed at time T, and let b be the deceleration; then

$$\frac{dv}{dt} = -b.$$

Apart from replacing a by $-b$ and the fact that the range of integration now starts from $t=T$ instead of $t=0$, the mathematics is very similar to what we did for the first phase:

$$\int_{T}^{t} \frac{dv}{dt}\, dt = \int_{T}^{t} -b\, dt$$

which gives

$$\int_{T}^{t} \frac{dv}{dt}\, dt = \int_{T}^{t} dv = [v]_{T}^{t} = v(t) - v(T).$$

As we are assuming b is constant,

$$\int_{T}^{t} -b\, dt = [-bt]_{T}^{t} = -b(t-T)$$

so

$$v(t) - v(T) = (-b)(t-T), \quad t \geq T,$$

or

$$v(t) = v(T) - b(t-T), \quad t \geq T.$$

As before, we integrate $v(t) = \dfrac{dx}{dt}$, to give

$$\int_{T}^{t} \frac{dx}{dt}\, dt = \int_{T}^{t} v(t)\, dt = \int_{T}^{t} (v(T) - b(t-T))\, dt.$$

Therefore,

$$[x(t)]_{T}^{t} = \left[v(T)t - \tfrac{1}{2}b(t-T)^2\right]_{T}^{t}$$

hence

$$x(t) - x(T) = v(T)(t-T) - \tfrac{1}{2}b(t-T)^2, \quad t \geq T,$$

or

$$x(t) = x(T) + v(T)(t-T) - \tfrac{1}{2}b(t-T)^2, \quad t \geq T.$$

We could have obtained this last result immediately from the previous general result, simply by noting that a is replaced by $-b$, t is replaced by $(t-T)$, and the "initial" position and velocity become $x(T)$ and $v(T)$. However, we have given the full derivation, for it provides another example of definite integration with a non zero lower limit.

For the acceleration phase we found $v(t)=at$ and $x(t)=\frac{1}{2}at^2$, so that $v(T)=aT$ and $x(T)=\frac{1}{2}aT^2$.

We can now combine the acceleration and braking phases to give

$$v(t) = \begin{cases} at, & \text{if } t \leq T \\ v(T) - b(t-T), & \text{if } t \geq T \end{cases}$$

and

$$x(t) = \begin{cases} \frac{1}{2}at^2, & \text{if } t \leq T \\ x(T) + v(T)(t-T) \\ \quad - \frac{1}{2}b(t-T)^2, & \text{if } t \geq T \end{cases}$$

Notice that the velocity is made up of two linear segments and the displacement from two quadratics, as shown in Figure 10.1.

Given values for the acceleration, a, the braking, b, and the time at which the brakes are applied, T, the position and velocity are determined. If we want to make sure that the dragster is brought to rest by the end a quarter mile track, then we need to solve $x(t)=0.5$ to find the total time of the run, t_R and then make sure that $v(t_R)=0$.

We will consider this in Problem 10.1.

WORKED EXAMPLE 10.2
SKY DIVING

In this example we will look at the mathematics behind the parachute jump application. In free-fall, the downward acceleration is due to gravity. Once the parachute is deployed, we will assume that the resistive force due to air resistance is proportional to v so that the deceleration is given by the linear expression $kv-g$, where k is some constant of proportionality.

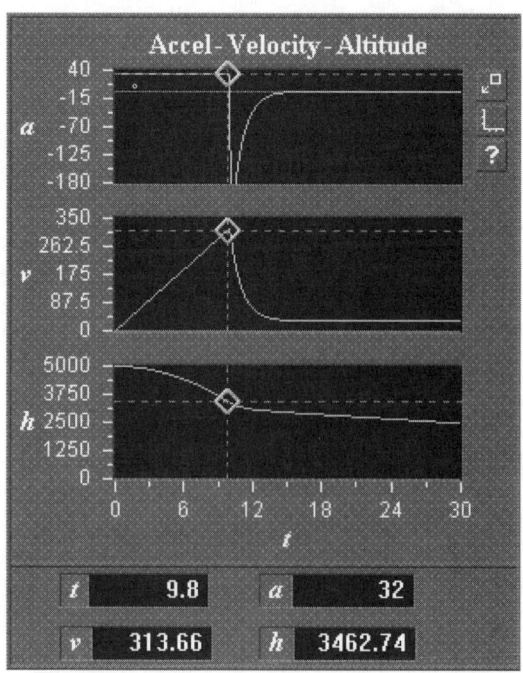

Figure 10.2. Acceleration, velocity, and displacement for a typical parachute jump.

In the software you can alter k using the slider, and change the time that the parachute opens to see what effect these have. You should obtain graphs like those shown in Figure 10.2.

The normal mathematical convention is to measure velocity as positive in the direction of increasing distance. But the Integral Calculus can handle any convention, so we will use a more natural one for this example. We will take $h(t)$ (in feet) to be the height of the parachutist, starting at $h=H$ at $t=0$, and we will take $v(t)$ to be the speed of descent (in feet per sec). As usual, we will take the acceleration due to gravity to be g (approximately 32 feet/sec/sec).

Since acceleration is given by the rate of change of velocity,

$$\frac{dv}{dt} = g.$$

We integrate this with respect to t over the range $t=0$ to t to obtain

$$\int_0^t \frac{dv}{dt} dt = \int_0^t g \, dt.$$

As g is constant, we obtain

$$v(t) = v(0) + gt.$$

Since v is measured downward we know $v(t) = -\frac{dh}{dt}$, and integration gives

$$\int_0^t \frac{dh}{dt} dt = -\int_0^t v(t) dt = -\int_0^t (v(0) + gt) dt.$$

$$\therefore \quad [h(t)]_0^t = -\left[v(0)t + \tfrac{1}{2}at^2\right]_0^t$$

$$\therefore h(t) - h(0) = -v(0)t - \tfrac{1}{2}at^2$$

$$\therefore \quad h(t) = h(0) - v(0)t - \tfrac{1}{2}gt^2.$$

In our case we start the stopwatch as the parachutist jumps, so that at $h(0)=H$ and $v(0)=0$, giving $h(t) = H - \tfrac{1}{2}gt^2$.

Next we need to consider the deceleration phase; from the time that the parachute opens, $t=T$, and let the speed at this time be

$v=v(T)$. We need to know the deceleration, and we will assume that from observations it has been found that the deceleration due to the parachute is proportional to the speed, and that this has to be combined with the continued effect of gravity. Suppose the constant of proportionality is k, so that acceleration is $g - kv(t)$.

The mathematics now is different from that of the first phase because not only is the acceleration no longer constant, but it also involves the function $v(t)$ which at this stage is unknown. We need to put the equation in a form in which we can use some standard results of integration. We have:

$$\frac{dv}{dt} = g - kv$$

therefore,

$$\frac{1}{g-kv}\frac{dv}{dt} = 1.$$

Integrating both sides with respect to t gives

$$\int_T^t \frac{dv}{g-kv} = \int_T^t dt$$

which we write as

$$\int_T^t \frac{-k\,dv}{g-kv} = \int_T^t -k\,dt$$

and integrate to give

$$[\ln(g-kv)]_T^t = [t]_T^t.$$

Now use the initial condition that $v=v(T)$ when the brakes are applied to give

$$\ln\left(\frac{g-kv(t)}{g-kv(T)}\right) = -k(t-T).$$

Finally, we can rearrange this as

$$v(t) = \frac{g}{k} - \left(\frac{g}{k} - v(T)\right)e^{-k(t-T)}.$$

Notice that as the exponential function $e^{-t} \to 0$ as $t \to \infty$, the velocity of the parachutist approaches g/k after sufficient time. This is called the terminal velocity. We can also see that if $v(T)<(g/k)$, then $v(t)$ will continue to increase, but if $v(T)>(g/k)$, then $v(t)$ will decrease.

Now we go on to find the height $h(t)$ as a function of t. Since $v(t) = -\frac{dh}{dt}$, we have:

$$\frac{dh}{dt} = -\frac{g}{k} + \left(\frac{g}{k} - v(T)\right)e^{-k(t-T)}.$$

Integrating both sides with respect to t gives

$$\int_T^t \frac{dh}{dt}\,dt = \int_T^t \left(-\frac{g}{k} + \left(\frac{g}{k} - v(T)\right)e^{-k(t-T)}\right)dt$$

and so

$$[h]_T^t = \left[-\frac{g}{k}t - \frac{\frac{g}{k} - v(T)}{k}e^{-k(t-T)}\right]_T^t.$$

Therefore,

$$h(t) = h(T) - \frac{g}{k}(t-T)$$
$$- \left(\frac{g}{k^2} - \frac{v(T)}{k}\right)\left(1 - e^{-k(t-T)}\right), \quad t \geq T.$$

In the free fall phase, we have already determined that $h(t) = H - \frac{1}{2}gt^2$, $t \leq T$; hence $h(T) = H - \frac{1}{2}gT^2$, which determines the height of the parachutist as a function of time. (See Figure 10.2 to see what the combined function looks like.)

PROBLEM 10.1
DRAGSTER ACTIVITIES

Name: _____

Date: _____

Section: _____

In this problem we return to the dragster example in Worked Example 10.1 and try to establish more details about the behavior of the vehicle.

1. What is 60 mph expressed in feet per second?

2. How long does it take to reach 60 mph at a constant acceleration of 8.8 ft/sec/sec?

3. How far (in feet) will the dragster have traveled when its speed hits 60 mph?

4. Suppose the deceleration from maximum is 24 ft/sec/sec. What is the minimum distance needed to get from 0-60 mph and then stop? How long does this take?

5. Suppose the acceleration is 16 ft/sec/sec and the braking deceleration is 24 ft/sec/sec. What is the maximum speed that can be reached on a quarter mile track if the dragster is to stop before the end of the track?

6. Suppose now that a more powerful engine is discovered which gives 25% better acceleration. How much must the braking be improved if the total run is to remain the same?

PROBLEM 10.2
SKY DIVER

Name: _____
Date: _____
Section: _____

In this problem we return to the parachute application and the mathematics described in Worked Example 10.2.

1. A parachutist jumps at 5000 feet and counts to 10 seconds before opening her 'chute. Taking g to be 32 ft/sec/sec and ignoring air resistance, how fast is she traveling at this time? What is her altitude at this point?

2. Assuming her terminal velocity is 20 ft/sec, find k. Assuming that she starts her jump as in (1) above, how long does it take her to descend and how fast is she traveling when she lands?

3. For a terminal velocity of 20 ft/sec, what is the lowest height she can open the parachute to make a landing at under 25 ft/sec?

4. Investigate the way that the minimum safe height changes if she starts her jump at higher altitudes.

NEW SITUATIONS

1. Dropping stones down a well

One way of estimating the depth of a well or mine shaft is to drop a stone and time the interval from releasing the stone to hearing the splash. Ignoring air resistance and the time for the sound to travel back up, the theory tells us that $x(t) = \frac{1}{2}gt^2$. Take $g=32$ ft/sec/sec, and use this to estimate the depth of a well when the time to the splash is:

   1 sec

   2 secs

   4 secs.

In each case work out the speed of the stone as it hits the well water.

Questions for discussion:

Is this a very accurate way of measuring the depth of wells?

How great a depth could reasonably be measured in this way?

To answer these questions you may need to think carefully about the physical effects important to this problem, not all of which have been introduced in the text! For example, is the speed of sound important? When will air resistance affect the results? You may also need to make some assumptions about other physical constants.

2. Lunar baseball

A fully trained NASA astronaut can hit a baseball straight up to a height of 50 feet on a terrestrial pitch. If he hits the ball just as hard on the moon (so it leaves the bat at the same speed as on earth), how high will the baseball go? Assume that lunar gravity is one tenth of its value on earth.

3. Braking distances

Imagine that you are helping to write a highway code for drivers, and you have been asked to set out safe braking distances.

You have consulted a physiologist who reports that reaction time may be safely assumed to be two-thirds of a second. This is the least time that elapses after the driver's brain notices an emergency situation and sends nerve signals to the leg muscles to hit the brake pedal.

You have consulted an engineer who tells you that on a dry road and with tires and brakes in good condition the maximum braking deceleration (but not a skid) is equivalent to $0.7g$.

Use these facts and the theory for straight line deceleration to construct a table of stopping distances made up of "thinking distance" and "braking distance". The entries in the table should be rounded to the nearest 5 feet and should cover speeds from 10 to 80 mph at 10 mph intervals.

Discussion point: what's the effect on stopping distance of doubling the speed?

4. More difficult problem: Parachuting - An alternative model for air resistance

In Worked Example 10.2 the parachute phase of the sky diver's descent was modeled by using a deceleration $g-kv$. Investigate a new mathematical model based on a deceleration given by $g-kv^2$.

Compare the expressions for the velocity and height after the parachute is opened.

5. Traffic calming

In many urban areas of our cities, attempts are being made to reduce the speed of the traffic by installing ramps in the road. Such ramps are designed so that the maximum speed of a car that passes over them is approximately 10 mph. If the car then accelerates uniformly and brakes when the next ramp becomes visible, it is possible to control the maximum speed that is achievable. If an average car can accelerate at 15 ft/sec/sec and brake at 20 ft/sec/sec, investigate the relationship between the distance between the ramps in terms of the initial and maximum velocities.

(*Hint:* You may find the result in Exercise 10.3 helpful.)

Chapter 11

Prerequisites

Before you study this material, you should be familiar with:

(1) The elements of differential and integral calculus (Modules 3, 5, 6, and 7).

(2) The concept that an integral can be thought of as the "area under the curve" (Module 6).

(3) The concept of definite integration and its applications (Module 9).

Objectives

In Chapter 9 Definite Integrals it was shown that the problem of finding the area under a curve can be solved by selecting a suitable antiderivative. In this Chapter we learn that "most" functions do not have a closed-form integral. This means that the integral of a typical function cannot be expressed in terms of simple functions, even if the function being integrated is itself composed of nothing but simple functions. Therefore, for many practical applications a reliable method of evaluating integrals is needed. Square counting as used in Chapter 4 is simple, but not very accurate; the next step toward a better method is Simpson's Rule.

We will introduce Simpson's Rule by showing that we can approximate any section of a smooth curve by a parabola and then, by integrating the analytic expression for the parabola, estimate the integral. We will show that just three data points are needed

Simpson's Rule

to define a parabolic section. Curves are built of sections and the area under each section is added to form an estimate for the integral under several sections.

Once we have an estimate, we need methods to determine the accuracy of the estimate. We will not attempt rigorous error analysis, but will show how to estimate errors from the data points used to define the parabolic sections.

Connections

The integral of a typical function arising in a practical application can seldom be expressed in terms of elementary functions. For practical purposes it is very important to have methods available for finding integrals numerically, because engineers, surveyors, business analysts, and so on all need hard figures. Before computers were widely available, mathematical models would sometimes be chosen on the basis of their solubility rather than their realism. Today, models do not need to be massaged to fit in with convenient mathematical functions, and computer packages will work with models that need to be chosen only for their realism. This theme is explored more fully in Chapter 16.

The software starts with two simple physical examples.

The first is solar heating. Power is a measure of the rate of transfer of energy, so the total energy input into a solar panel is the integral of the power input. The temperature rise is directly proportional to the energy input.

The second software example is again that of dragsters, as analyzed in Chapter 10. There, it was assumed that the acceleration was constant (until braking). This isn't true in fact, for the power developed by an engine depends on its rpm. (Rpm stands for revolutions per minute and measures the speed of rotation of the engine.) The characteristic curve of power versus rpm will be determined from bench tests on the engine, and is not likely to be easily describable by any elementary mathematical function. If we want to use bench tests to predict the dragster's performance, we will need a method of integration that applies to measured data rather than only to elementary mathematical functions.

Fitting a Parabola to Data

In any small interval, a smooth function can be approximated by a parabola (see Figure 11.1). Therefore, we divide the range of integration [a, b] for the integral

$$\int_a^b f(x)dx$$

into panels and approximate the integrand in each panel by a suitable parabola.

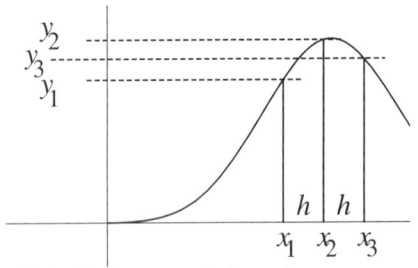

Figure 11.1. Fitting a parabola to data.

Remember, the general form a function whose graph is a parabola is $ax^2 + bx + c$.

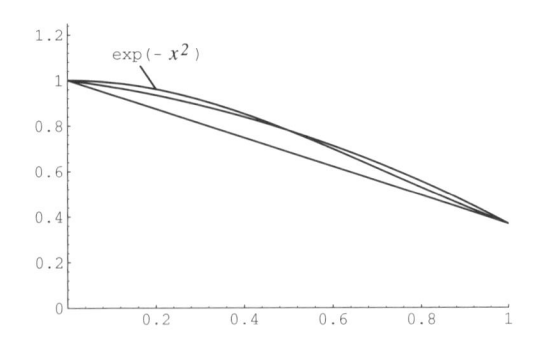

Figure 11.2. A linear and quadratic approximation to $\exp(-x^2)$ over [0,1].

In Chapter 9 numerical methods were introduced, and the trapezoidal rule was discussed. Recall that in effect the trapezoidal rule approximates the integrand (the function to be integrated) by a linear function for each panel of the integration. The reason for now choosing a parabola is that it is a simple curve that ought to be "better" than a straight line at approximating curves, as illustrated in Figure 11.2. It was easy to find the area of a trapezoid, and it is only slightly more complicated to find the area under a parabola (because we can integrate functions like $ax^2 + bx + c$).

There are several ways to find a parabola that fits given data. Note first that the formula for the parabola has three unknown parameters a, b, and c. To determine suitable values for a, b, and c, it is therefore necessary to fit the data at three points, since this will give three equations for three unknowns. Suppose these three points are (x_1, y_1), (x_2, y_2), and (x_3, y_3). This procedure gives the three equations:

$$\left.\begin{aligned} ax_1^2 + bx_1 + c &= y_1 \\ ax_2^2 + bx_2 + c &= y_2 \\ ax_3^2 + bx_3 + c &= y_3 \end{aligned}\right\} \quad (11.1)$$

These can be solved (a little tediously) by hand or by "brute force" using computer algebra. There are, however, more elegant ways that are capable of being extended to more general cases; these are the methods of (1) iterated polynomials and (2) Lagrange polynomials. The next two sections give the details, and then we present Simpson's Rule itself.

Example 11.1

Fit a quadratic through the points (0, 1), (1, 3), (2.5, 13.5).

Applying equations (11.1) to $ax^2 + bx + c$ for these points gives

$$0a + 0b + c = 1$$
$$a + b + c = 3$$
$$6.25a + 2.5b + c = 13.5$$

The first equation gives c immediately, leaving

$$a + b = 2$$
$$6.25a + 2.5b = 12.5$$

Multiply the first of this new pair of equations by 2.5 and subtract from the last equation to give

$$3.75a = 7.5.$$

Therefore, a=2 and b=0, and the required function is $2x^2 + 1$.

> **Exercise 11.1**
>
> **Fitting a parabola to data**
> Solve equations (11.1) using pencil and paper, and verify your results using a computer algebra system.

The Method of Iterated Polynomials

Let us begin by looking at one panel, which we will take to be the interval $[x_1, x_3]$ with x_2 as the midpoint. Let h be the "half-panel size" so that

$$h = (x_2 - x_1) = (x_3 - x_2) \quad (11.2)$$

We want to find a function $f(x)$ whose values match the data at given values of x, that is, $y_1 = f(x_1)$, $y_2 = f(x_2)$, $y_3 = f(x_3)$.

Define $p_0(x) = y_1$. This has degree zero since its value is a constant, independent of x. It fits the point (x_1, y_1) since trivially $p_0(x_1) = y_1$.

Now define $p_1(x) = p_0(x) - A(x - x_1)$ where A is a constant to be determined. This is a polynomial of degree 1. It still fits the point (x_1, y_1) because the additional term is zero for $x = x_1$. We choose A to make it fit (x_2, y_2); that is, we want $p_1(x_2) = y_2$. This gives

$$A = \frac{y_2 - y_1}{x_2 - x_1} = \frac{y_2 - y_1}{h}.$$

Next define:

$$p_2(x) = p_1(x) + B(x - x_1)(x - x_2),$$

where B is another constant. This is a polynomial of degree 2. The function still fits the points (x_1, y_1) and (x_2, y_2) because the additional term is zero for $x = x_1$ and $x = x_2$. We choose B to fit (x_3, y_3), that is, $p_2(x_3) = y_3$. This gives

$$B = \frac{y_3 - y_1 - A(x_3 - x_1)}{(x_3 - x_1)(x_3 - x_2)} = \frac{y_3 - 2y_2 + y_1}{2h^2},$$

and therefore $p_2(x)$ is what we wanted, a polynomial of degree 2 (i.e., a parabola) that fits three points on the curve $y = f(x)$, that is, $p_2(x_1) = y_1$, $p_2(x_2) = y_2$, and $p_2(x_3) = y_3$. Therefore, $p_2(x)$ is a polynomial of degree 2 that agrees with $f(x)$ at the points given.

We have

$$p_2(x) = y_1 + A(x - x_1) + B(x - x_1)(x - x_2) \quad (11.3)$$

where $A = \dfrac{y_2 - y_1}{h}$

and
$$B = \dfrac{y_3 - 2y_2 + y_1}{2h^2}.$$

> **Exercise 11.2**
>
> **Verifying the polynomial fit**
> Verify that the polynomial defined by (11.3) goes through the points (x_1, y_1), (x_2, y_2), and (x_3, y_3). Check your results using a computer algebra system.

To derive Simpson's Rule, let us approximate $f(x)$ by $p_2(x)$, that is
$$f(x) \approx p_2(x).$$
Then
$$\int_{x_1}^{x_3} f(x)dx \approx \int_{x_1}^{x_3} p_2(x)dx.$$

Let $S = \int_{x_1}^{x_3} p_2(x)dx$, then

$$S = y_1 \int_{x_1}^{x_3} dx + A \int_{x_1}^{x_3} (x - x_1)dx +$$
$$B \int_{x_1}^{x_3} (x - x_1)(x - x_2)dx \quad (11.4)$$

The three integrals in (11.4) are quite easy to work out using the substitution $z = (x - x_1)$. This gives:

$$\int_{x_1}^{x_3} dx = \int_0^{2h} dz = 2h$$

$$\int_{x_1}^{x_3} (x - x_1)dx = \int_0^{2h} z\,dz = 2h^2$$

$$\int_{x_1}^{x_3} (x - x_1)(x - x_2)dx = \int_0^{2h} z(z - h)dz = \tfrac{2}{3}h^3$$

Combining these results gives
$$S = 2hy_1 + 2h^2 A + \tfrac{2}{3}h^3 B,$$
which simplifies to
$$S = \tfrac{1}{3}h(y_1 + 4y_2 + y_3). \quad (11.5)$$

> **Exercise 11.3**
>
> **Simpson's Rule from iterated polynomials**
> Verify the integrals used to obtain (11.5) using pencil and paper, and check your results using a computer algebra system.

The Method of Lagrange Polynomials

Although a little harder to understand than the iterated polynomial method, the Lagrange polynomial method is a powerful technique that can be extended to derive other, more elaborate formulas.

We start as before by looking at just one panel on the interval $[x_1, x_3]$ with x_2 as the midpoint. Let h be the "half-panel size" as in (11.2), and define $y_1 = f(x_1)$, $y_2 = f(x_2)$, $y_3 = f(x_3)$ as before.

Now define three functions:

$$\left.\begin{aligned} L_1(x) &= \dfrac{(x - x_2)(x - x_3)}{(x_1 - x_2)(x_1 - x_3)} \\ L_2(x) &= \dfrac{(x - x_1)(x - x_3)}{(x_2 - x_1)(x_2 - x_3)} \\ L_3(x) &= \dfrac{(x - x_1)(x - x_2)}{(x_3 - x_1)(x_3 - x_2)} \end{aligned}\right\} \quad (11.6)$$

Notice too that

a. symmetry is used in labeling and defining these functions: $L_i(x)$ is defined so that x_i does not appear in the numerator, and x_i is the first term in each of the brackets in the denominator;

b. each of these functions is a parabola because it is a quadratic in x; and

c. each is zero at two of the points x_1, x_2, x_3, and 1 at the other. In fact, $L_1(x_1)=1$, $L_1(x_2)=0$, and $L_1(x_3)=0$, and similarly for $L_2(x)$ and $L_3(x)$.

Now consider the function defined by
$$Q(x) = y_1 L_1(x) + y_2 L_2(x) + y_3 L_3(x) \quad (11.7)$$

$Q(x)$ is a parabola, because each $L_i(x)$ is quadratic in x and so $Q(x)$ must also be quadratic. From the properties of each $L_i(x)$ it is also true that $Q(x_1)=y_1$, $Q(x_2)=y_2$, and $Q(x_3)=y_3$.

$Q(x)$ is therefore what we wanted, a polynomial of degree 2 (i.e., a parabola) that fits three points on the curve $y=f(x)$.

To derive Simpson's Rule, let us approximate $f(x)$ by $Q(x)$; that is,
$$f(x) \approx Q(x).$$

Then
$$\int_{x_1}^{x_3} f(x)dx \approx \int_{x_1}^{x_3} Q(x)dx.$$

Let $S = \int_{x_1}^{x_3} Q(x)dx$, then
$$S = y_1 \int_{x_1}^{x_3} L_1(x)dx + y_2 \int_{x_1}^{x_3} L_2(x)dx +$$
$$y_3 \int_{x_1}^{x_3} L_3(x)dx \quad (11.8)$$

The three integrals in (11.8) are quite easy to work out using the substitution $z = (x - x_1)$. This gives

$$\int_{x_1}^{x_3} L_1(x)dx = \int_0^{2h} \frac{(z-h)(z-2h)}{2h^2} dz = \tfrac{1}{3}h,$$

$$\int_{x_1}^{x_3} L_2(x)dx = \int_0^{2h} \frac{z(z-2h)}{-h^2} dz = \tfrac{4}{3}h,$$

$$\int_{x_1}^{x_3} L_3(x)dx = \int_0^{2h} \frac{z(z-h)}{2h^2} dz = \tfrac{1}{3}h.$$

Combining these results gives, as before,
$$S = \tfrac{1}{3}h(y_1 + 4y_2 + y_3) \quad (11.9)$$

> **Exercise 11.4**
>
> **Simpson's Rule from Lagrange polynomials**
> Verify the integrals used to obtain (11.9) using pencil and paper, and check your results using a computer algebra system.

Simpson's Rule for Several Panels

The two results (11.5) and (11.9) are identical, though derived by different methods. This should be no surprise, because there can only be one parabola that fits the three data points. In fact, the functions $p_2(x)$ defined in equation (11.3) and $Q(x)$ defined in (11.7) must be the same. They only appear different because their terms have been arranged in a different order.

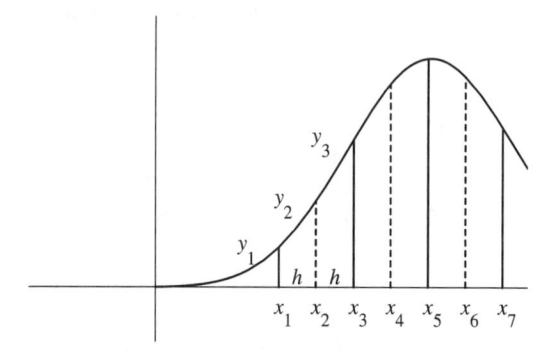

Figure 11.3. Simpson's Rule for several panels.

To use Simpson's Rule in practice, the integration interval is divided into several panels. There is no mathematical requirement that these panels should have the same width, but the most commonly used form of Simpson's Rule uses equal-sized panels, with h as the half-panel size. Simpson's Rule for several panels is obtained by adding the results for each panel. For example, with three panels as shown in Figure 11.3, we get

$$\int_{x_1}^{x_7} f(x)dx \cong \tfrac{1}{3}h\bigl[(y_1 + 4y_2 + y_3) + (y_3 + 4y_4 + y_5) + (y_5 + 4y_6 + y_7)\bigr]$$

$$= \tfrac{1}{3}h\bigl[y_1 + 4y_2 + 2y_3 + 4y_4 + 2y_5 + 4y_6 + y_7\bigr].$$

The extension to N panels is evidently

$$\int_{x_1}^{x_7} f(x)dx \cong \tfrac{1}{3}h\bigl[y_1 + 4y_2 + 2y_3 + 4y_4 + \ldots + 4y_{2n} + 2y_{2n+1} + \ldots + 4y_{2N} + y_{2N+1}\bigr] \quad (11.10)$$

Example 11.2
Use the software to try out Simpson's Rule for the following integral whose result is known:

$$\int_0^1 \frac{4dx}{1+x^2} = 4[\arctan(x)]_0^1 = \pi = 3.14159265\ldots.$$

Within a factor of 2, how many panels are needed to obtain a result correct to eight significant figures?

Draw up a table recording the number of panels n, the value of h, and the estimate S for the integral.

n	h	S
2	0.25	3.141568621
4	0.125	3.141592503
8	0.063	3.141592642
16	0.031	3.141592676
32	0.016	3.141592659

When $n=32$ observe that eight significant figures have remained unchanged from the previous line. We conclude that 16 panels are sufficient to obtain eight significant figures in this case.

Error Estimation

With any numerical method it is just as important to estimate the error of any result as to obtain the result itself. Arguably, the error estimate is even more important than the result, since a wrong result may lead to false conclusions, wasted time and effort, or even endangerment to life if an engineering project depends on the results.

Detailed error analysis is beyond the scope of this chapter, but there are simple ways of estimating errors that can be used without relying on advanced theory.

The obvious first step with any numerical method that uses a finite step size (like the half-panel size h in Simpson's Rule) is to repeat the calculation with new values. For Simpson's Rule, intuition suggests that the more panels are used, the better a parabola will fit the data in each panel, and so reducing h might be expected to increase the accuracy of the result. A good rule of thumb is successively to double the number of panels (thus successively halving h). In Example 11.2 the number of panels was doubled until the result remained unchanged to the required number of significant figures. It should be noted, however, that this rule does not guarantee that the apparent accuracy is really achieved. It will be achieved under most circumstances, but if the number of figures is close to the maximum accuracy of the computer, or the function or its derivatives are unusually large, the results must be treated with caution.

If the results of applying Simpson's Rule with $h = h_0$ and $h = \tfrac{1}{2}h_0$ give answers that agree to the required number of figures, then apart from exceptional cases it is reasonable to accept the results.

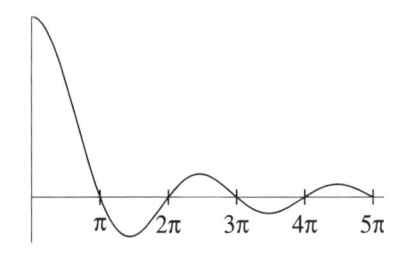

Figure 11.4. $y = \dfrac{\sin(x)}{x}$.

An exceptional case (admittedly contrived) is $\int_{\pi}^{5\pi} \frac{\sin(x)}{x} dx$ (see Figure 11.4). Simpson's Rule gives zero for both one panel ($h=2\pi$) and two panels ($h=\pi$). Special cases like this must be eliminated by inspection. Smaller values of h will not give zero, and normal convergence will take place.

Exercise 11.5

Estimating an integral

Use the software to evaluate $\int_{\pi}^{5\pi} \frac{\sin(x)}{x} dx$ to four significant figures.

There is an unspoken assumption in the above rule of thumb that as h is successively halved, the accuracy of the results will continue to improve. It would be useful to know if this is true, and if not, just how small is it sensible to take h? If a computer or calculator were to have "infinite precision" (that is, be able to store as many significant figures as you ever wanted), then it would indeed be true that Simpson's Rule could be used to obtain results to any desired accuracy. However, the floating-point processors installed in computers have a finite precision. (Computer algebra systems sometimes get around this by using software that can deliver any required precision, limited only by the capacity of the computer's memory and the time you are prepared to wait for the answer!) How can the user tell when the best precision has been achieved? It would also be useful to be able to predict how to obtain h to obtain the required accuracy. Numerical analysis provides an answer to these questions, but once again it is possible to see how to proceed (in most cases) without resorting to advanced mathematics.

Let S_n be the result of applying Simpson's Rule using n panels. Suppose that the exact result is known, say, equal to I. Define the error to be

$$E_n = S_n - I \qquad (11.11)$$

The method is to plot $-\log_{10}|E_n|$ against $-\log_2 h$. Logs to base 10 are convenient for the vertical scale, as they are closely related to the number of leading zeros in the error, and the data points move higher up the vertical scale as the accuracy improves. Because of the use of $-\log(h)$, halving h takes the horizontal coordinate of each data point $\log(2)$ to the right [since $-\log(\frac{1}{2}h) = -\log(h) + \log(2)$]. After the interval has been halved m times, we have $n = 2^m$ and $h = 2^{-m} h_0$. Therefore,

$$-\log_2 h = m - \log_2(h_0).$$

(Recall that $\log_{10}(x) = \log_{10}(2)\log_2(x)$, and $\log_2 2^m = m$). In Figure 11.5, each horizontal space between data points represents a halving of h, and each vertical division represents an improvement in accuracy by a factor of 10.

Example 11.3

Plot a log-log error graph for Simpson's Rule with the integral

$$\frac{2}{\sqrt{\pi}} \int_0^1 \exp(-t^2) dt = \mathrm{erf}(1) = 0.8427007929\ldots$$

Draw up a table recording the number of panels n, the value of h, the estimate S_n for the integral, and $-\log|E_n|$.

n	h	S_n	$-\log_{10}(|S_n - \mathrm{erf}(1)|)$
2	0.25	0.8427360513…	4.45
4	0.125	0.8427030358…	5.65
8	0.063	0.8427009335…	6.85
16	0.032	0.8427008017…	8.06

The resulting graph is shown in Figure 11.5.

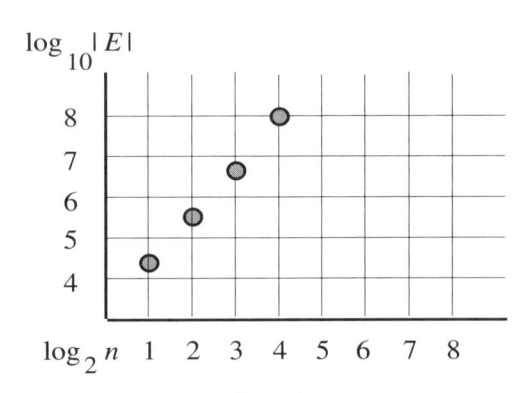

Figure 11.5. $-\log_{10}(|\text{error}|)$ *against* $\log_2(n)$.

It turns out (and can be proved analytically) that if the effect of limited computer accuracy is ignored, the points on this plot will tend to lie on a straight line for small enough h. The slope of this line will depend on the method (Simpson's Rule in this case) and on properties of the integrand. The trend of the line can be used in principle to estimate the panel size needed to obtain a desired accuracy. In practice, limited computer accuracy will interfere with this result, and the points on the plot will depart from the line. In particular, the trend when h gets really small (and n gets large) is for the accuracy to decrease. Figure 11.5 shows a situation where accuracy has consistently improved and the slope of the trend is positive. The effect of limited machine precision will eventually be to reduce the slope. For very large n a downward trend with negative slope may be introduced. Therefore, it is to be expected that the graph will show a maximum. Such a point will represent the best accuracy obtainable by using Simpson's Rule on a given computer. Figure 11.6 illustrates this situation. It must be noted that errors due to limited computer accuracy seldom behave smoothly, and the overall trends described will usually have erratic effects superimposed. Futhermore, the behavior at large n is highly sensitive to the way that computer hardware and software handle floating point numbers, and results obtained on different computers will almost certainly differ from those shown here.

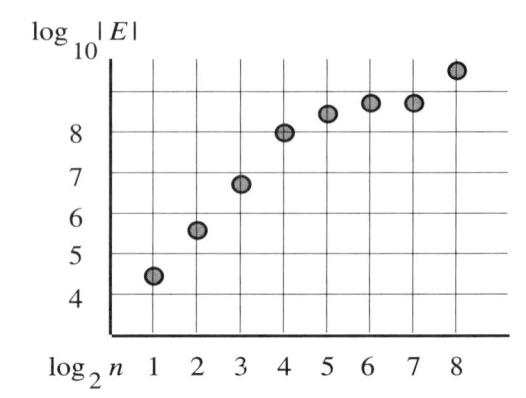

Figure 11.6. $-\log_{10}(|\text{error}|)$ *against* $\log_2(n)$.

Exercise 11.6

Limits to accuracy

Extend the table of results shown in Example 11.3 to verify Figure 11.6.

To explain the theory so far a known integral has been used, but the whole point of numerical methods like Simpson's Rule is that they can be applied in cases where the result is not given by previously known mathematical quantities. The log-log error graph method can be adapted, simply by treating, at each stage, the result for the largest number of panels as if it were the correct result and using it to estimate the error. Thus, letting S_n be the result using n panels, the error is estimated to be

$$\hat{E}_n = S_n - S_{2n} \quad (11.12)$$

instead of $S_n - I$ as in (11.11). This technique will be demonstrated in Worked Example 11.1.

WORKED EXAMPLE 11.1
AREA ENCLOSED BY A QUARTIC CURVE

An ellipse with its center at the origin and semi-axes a and b is given by

$$\frac{x^2}{a^2} + \frac{y^2}{b^2} = 1.$$

The area enclosed by such an ellipse is πab (see Chapter 6), a result that can be obtained exactly. For other curves that may seem similar, exact formulas for area may not exist. Consider, for example, the quartic curve

$$\frac{x^4}{a^4} + \frac{y^4}{b^4} = 1.$$

Using symmetry, we see that the area for given values of a and b is given by

$$A(a,b) = \frac{4b}{a}\int_0^a (a^4 - x^4)^{\frac{1}{4}} dx$$

which with a simple change of variable $x=az$ can be shown to equal

$$A(a,b) = ab\int_0^1 4(1-z^4)^{\frac{1}{4}} dz$$

for which there is no closed-form solution. Now let

$$I = \int_0^1 4(1-z^4)^{\frac{1}{4}} dz$$

and investigate the value of I using Simpson's Rule.
Proceeding as in Example 11.3 except that the final column is calculated using equation (11.12) gives

n	h	S_n	$-\log_{10}(|S_n - S_{2n}|)$
2	0.25	3.5337378...	1.00
4	0.125	3.6350754...	1.37
8	0.063	3.6774778...	1.75
16	0.032	3.6952644...	2.13
32	0.016	3.7027341...	2.50
64	0.008	3.7058730...	2.88

The resulting graph is shown in Figure 11.7.

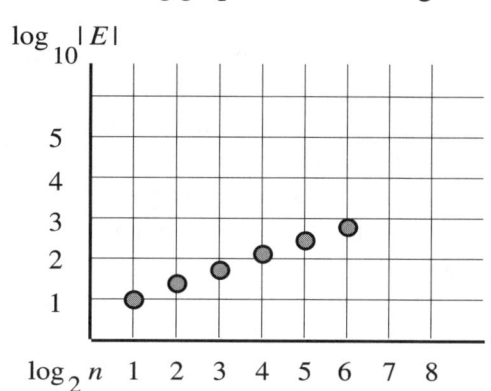

Figure 11.7. $-\log_{10}(|\text{est error}|)$ *against* $\log_2(n)$.

This gives sufficient results to estimate how many panels are needed for a given accuracy by extrapolating the trend of the graph. For five significant figures, for example, the error must be less than 5×10^{-5}; therefore, we require $-\log_{10}(|\text{error}|) > 4.3$. Note that the error column is fitted very well by

$$-\log_{10}|E| = 0.623 + 1.249(\log_{10} 2)(\log_2 n).$$

Therefore, to achieve five significant figures requires $\log_2 n > 9.8$, indicating that 2^{10} panels are needed. As a check, carry out Simpson's Rule for 2^9, 2^{10}, and 2^{11} panels, giving the following results:

$\log_2 n$	S_n	$-\log_{10}(|S_n - S_{2n}|)$
9	3.7079801...	4.01
10	3.7080782...	4.38
11	3.7081194...	

We see that the results for 2^9 and 2^{10} panels do not agree to 5 figures, but the results for 2^{10} and 2^{11} panels do. The predictions for accuracy are confirmed, and the area enclosed by the quartic curve is therefore $3.7081ab$ to five significant figures.

We now ask, what is the maximum accuracy that can be achieved in this Worked Example? Most modern personal systems support about 12 - 18 figures with numerical processing (but note that if computer algebra is used, in some systems the number of digits may be set arbitrarily). To reach 18 figure accuracy in our result here would require $-\log_{10}(|\text{error}|) > 17.3$. Using the same extrapolation as before, we would need $\log_2 n > 44.3$, indicating that 2^{45} panels would be needed! This gives $h \cong 10^{-14}$ and is so small that we would expect rounding error to be very significant. Also, if it takes 1 sec to compute the result for 1000 panels, then 2^{45} panels would take over 1000 years! Even 9 figure accuracy calls for 2^{21} panels (over 3 weeks on the same assumptions).

We must conclude that with a numerical method like Simpson's Rule, it may not be possible to achieve the full number of significant figures of which the hardware is capable. A balance must be struck between the errors due to the numerical approximation, the errors due to limited machine accuracy, and the time taken.

In the problem under consideration, it would take an impracticably large number of panels to show the onset of machine errors, but as we have seen illustrated in Figure 11.6, in some cases the log-log error plots do show the effects of limited machine accuracy with a much lower number of panels. Why should the problem in Example 11.3, for instance, show 9 figure accuracy with 2^8 panels, whereas in this problem 2^{21} are needed? The difference lies in the slope of the error graph. The error column for 2 - 16 panels in Example 11.3 is fitted very well by

$$-\log_{10}|E| = 3.247 + 3.997(\log_{10} 2)(\log_2 n).$$

Notice the coefficient 3.997, which implies that the error behaves very nearly like h^4. This is the usual behavior for Simpson's Rule when the integrand is well-behaved over the integration interval. For the integral I we have found that the error is behaving very nearly like $h^{1.25}$. It is this lower power which is slowing the convergence of the results, and it is due to the fact that the integrand in

$$I = \int_0^1 4(1-z^4)^{\frac{1}{4}} dz$$

has a singular derivative at $z = 1$.

PROBLEM 11.1 USING
SIMPSON'S RULE

Name: _____

Date: _____

Section: _____

The following integrals all have closed-form solutions. Find these either by pencil and paper or by using a CAS. Then investigate how well Simpson's Rule works for each of them, by plotting two log-log error graphs:

(1) $-\log_{10}(|E_n|)$ against $\log_2(n)$ as in Example 11.3, Figure 11.5, and

(2) $-\log_{10}(|S_n - S_{2n}|)$ against $\log_2(n)$ as in Worked Example 11.1, Figure 11.7.

Aim for eight significant figures in your most accurate estimates, but do not continue to increase n beyond the point where computation times are reasonable.

(i) $S = \int_0^1 4(1-z^2)^{\frac{1}{2}} dz$

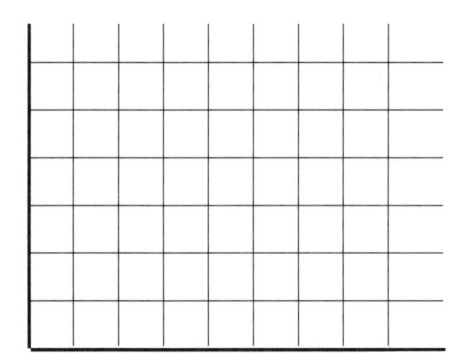

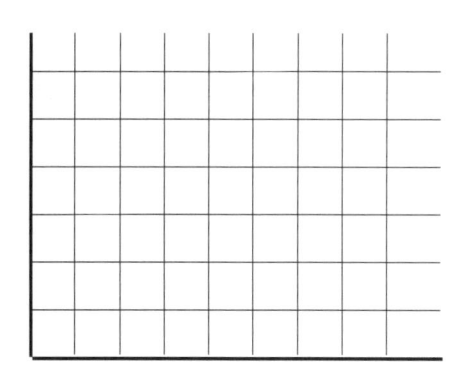

(ii) $S = \int_0^1 (1-x)^{\frac{1}{2}} dx$

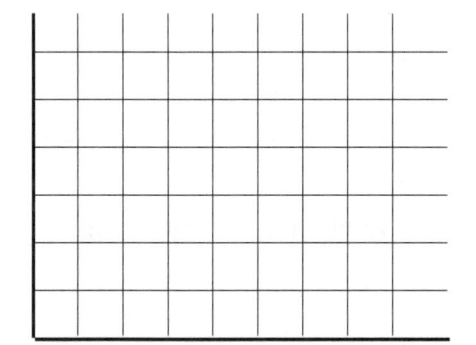

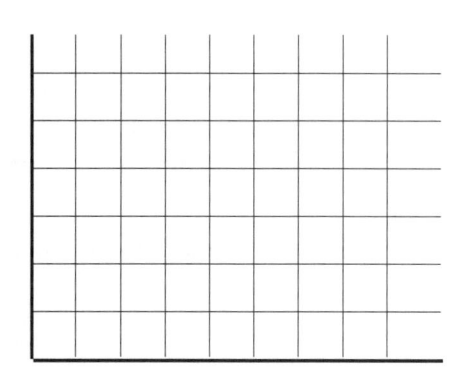

PROBLEM 11.1
USING SIMPSON'S RULE
(Continued)

(iii) $S = \int_{1}^{5} (2x^2 - 3x + 1) dx$

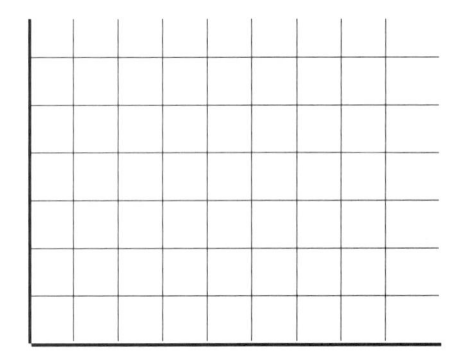

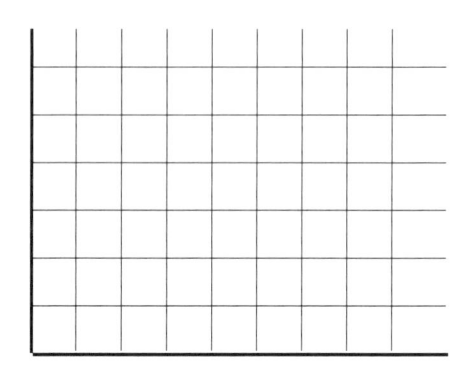

What is special about this integral, given that Simpson's Rule is being used?

(iv) $S = \int_{-2}^{2} (x^3 - 2x^2 + 3x - 1) dx$

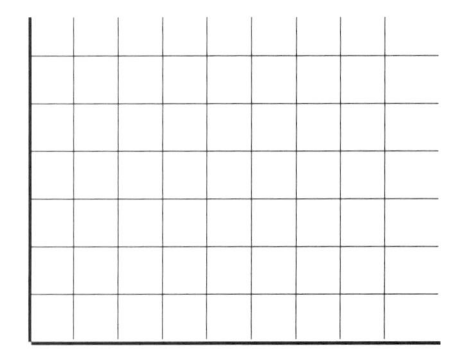

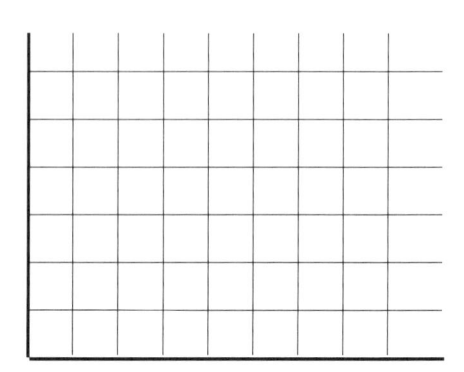

What do you notice about these results? Why do you think they are happening?

PROBLEM 11.2
FURTHER PROBLEMS

Name: _____
Date: _____
Section: _____

Problem 11.2

(i) Attempt to evaluate $\int_0^{10} \frac{dx}{(x-5)^2 + 0.001}$ using Simpson's Rule. First, sketch the function or use software to plot it. Why is it awkward? What is your best estimate of the result? Find the exact answer by pencil and paper or using a CAS.

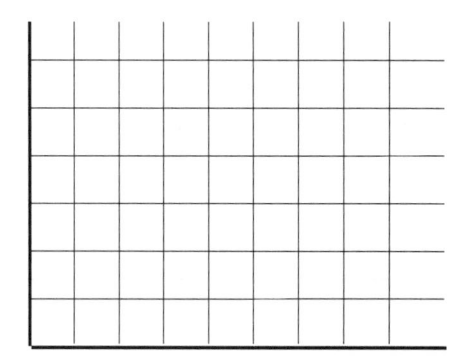

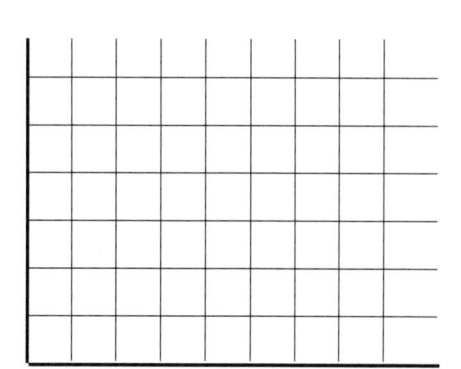

(ii) Evaluate $\int_1^2 x^2 e^{-x^2} dx$ using Simpson's Rule. First, sketch the function or use software to plot it. Find (to within a factor of 2) the largest value of h that will give the result to six significant figures. There is no exact answer in this case.

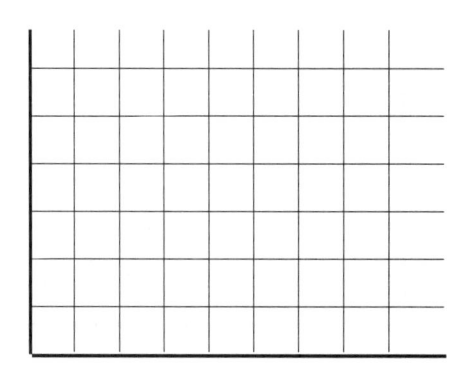

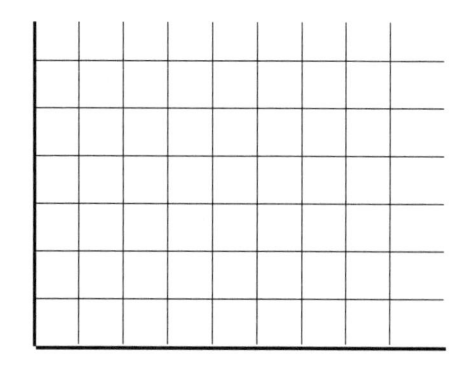

PROBLEM 11.2
FURTHER PROBLEMS
(Continued)

(iii) Evaluate $\int_1^2 x^{1/2} e^{-x^2} dx$ using Simpson's Rule. First, sketch the function or use software to plot it. Find (to within a factor of 2) the largest value of h that will give the result to six significant figures. There is no exact answer in this case.

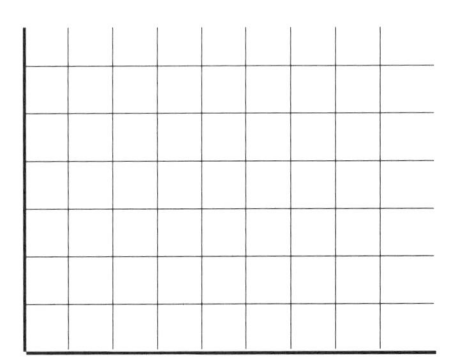

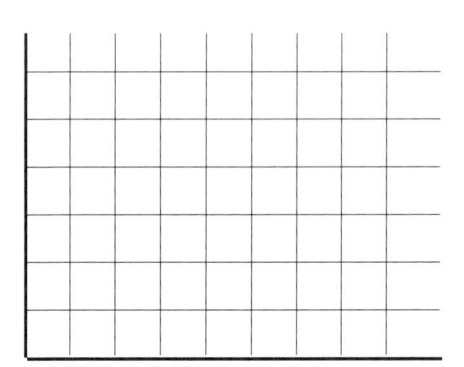

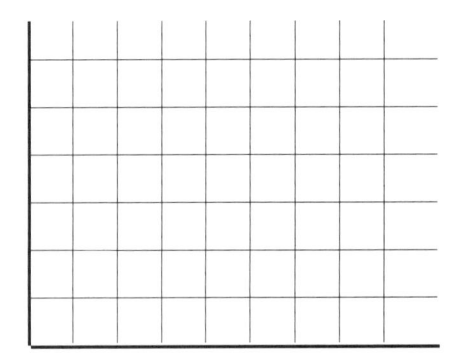

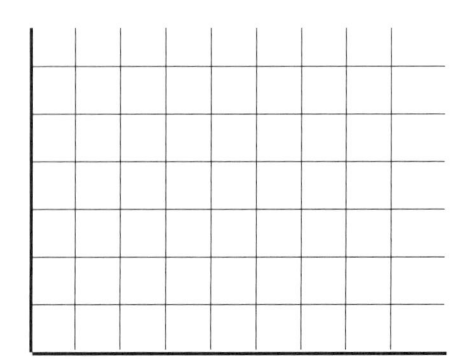

(iv) Evaluate $\int_0^\pi \frac{\sin(x)}{x} dx$ using Simpson's Rule. First, sketch the function or use software to plot it. Find (to within a factor of 2) the largest value of h that will give the result to six significant figures. There is no exact answer in this case.

NEW SITUATIONS

1. Order of convergence

In earlier chapters the trapezoidal rule for an integral $\int_a^b f(x)dx$ was given as

$$T_N(a,b) \cong \frac{h}{2}(y_0 + 2y_1 + 2y_2 + \ldots 2y_{N-1} + y_N)$$

where $y_i = f(x_i)$. In New Situations at the end of Chapter 8, the Mean Value Theorem was used to obtain an error estimate for the trapezoidal rule as:

$$2\frac{(b-a)^3}{N^2} \operatorname*{Sup}_{a \leq x \leq b} |f''(x)|.$$

The feature of interest in this result is the way that error depends on N. Notice that with N as defined here, the number of panels as used in Simpson's Rule is $\frac{1}{2}N$, but h has the same meaning.

The log-log error plotting techniques introduced in this chapter can also be applied to the trapezoidal rule. Use the software to help you do this for the same integral as in Example 11.2. You should observe that the trend of the error plot is less steep than for Simpson's Rule, which means that the accuracy of the result increases less rapidly with increasing N than is the case with Simpson's Rule.

We can explain the trend as follows. Because $h = \frac{b-a}{N}$, we can write the error estimate for the trapezoidal rule in the form $E \cong Ch^2$.
Therefore,

$$-\log(E) \cong -\log(C) + 2(-\log(h))$$

and the plot of -log(E) against -log(h) should be approximately a line of slope 2.

Verify that this is approximately true for your results. (Be careful to allow for appropriate units of measurement along each axis.)

In general for any integration rule that has error given by $E \cong Dh^k$, the log-log error plot should have slope k, and this is called the *order of convergence* of the method. You should find that for Simpson's Rule, $k \cong 4$ is consistent with results.

2. Method of undetermined coefficients

It may have occurred to you that, despite the length and complexity of the methods that were used to derive Simpson's Rule, the final result for one panel

$$S = \tfrac{1}{3}h(y_1 + 4y_2 + y_3)$$

is delightfully simple. Could it have been obtained in another way?

Any method that estimates $\int_a^b f(x)dx$ by fitting a polynomial through data points (x_1,y_1), (x_2,y_2), ..., (x_i,y_i) ... (x_{p+1},y_{p+1}) will generate a result of form

$$R = c_1y_1 + c_2y_2 + \ldots + c_iy_i + \ldots + c_{p+1}y_{p+1}$$

because a polynomial of degree p in x has $p+1$ coefficients. Furthermore, the result R must be exact, *not* approximate, for all functions $f(x)$ which are polynomials of degree p or lower. It is always possible to choose $p+1$ independent polynomials of degree p or lower, and so obtain $p+1$ equations for the unknown coefficients $c_1, c_2 \ldots c_{p+1}$.

Applying this line of argument to Simpson's rule, take $p = 2$, $x_1 = 0$, $x_2 = h$, and $x_3 = 2h$,

The second application, sporting records, provides another elementary example of sequences. A quick look at the history of individual events gives a variety of different types of sequences.

For example, the Olympic Pole Vault records for the period 1900 to 1912 are shown in Table 12.1.

Year	Height
1900	3.34m
1904	3.53m
1908	3.73m
1912	3.92m

Table 12.1. Olympic Pole Vault records.

Clearly, this is an increasing sequence; but will it ever reach a limit? Alternatively, recent world record times in the 1500 meters sprint are given in Table 12.2.

Year	Athlete	Time(m)
1967	Jim Ryan (USA)	3:33.10
1974	Filbert Bayi (Tan)	3:32.20
1980	Steve Ovett (GBR)	3:31.36
1983	Sydney Maree (USA)	3:31.24
1985	Said Aouita (Mar)	3:29.46
1992	Nouri Morceli (Alg)	3:27.37

Table 12.2. World records in the 1500 meters sprint.

This sequence is always decreasing but can never become negative; it is bounded below and so must converge.

In this chapter we will look at the convergence of sequences and see how the results obtained can be used to investigate the convergence of infinite series.

Limits of Sequences and Series

In Chapter 2 we looked at the convergence of simple sequences. A sequence is an ordered set of values a_n that is defined for positive integer values of n. For example,

$$1, 2, 3, 4, 5, 6, 7, \ldots$$

is a sequence with $a_n = n$, and

$$1, \tfrac{1}{4}, \tfrac{1}{9}, \tfrac{1}{16}, \tfrac{1}{25}, \ldots$$

is a sequence with $a_n = \dfrac{1}{n^2}$.

A *series* is formed by adding together the values of a sequence. For example,

$$1 + 2 + 3 + 4 + 5 + 6 + 7 + \ldots$$

$$1 + \tfrac{1}{4} + \tfrac{1}{9} + \tfrac{1}{16} + \tfrac{1}{25} + \ldots$$

are examples of series. If a sequence is defined for all integer values of n, (so that there are infinitely many members of the sequence), then the corresponding series is called an *infinite series*.

The question arises: Does a given sequence have a limit? In other words, does a_n approach a unique value a as n gets larger and larger? If there is such a value, it is called the "limit as n tends to infinity". This is written $a_n \to a$ as $n \to \infty$. It is obvious that the sequence $a_n = n$ does not have a limit: $a_n \to \infty$ as $n \to \infty$, whereas the sequence given by $a_n = \dfrac{1}{n^2}$ does have a limit: $a_n \to 0$ as $n \to \infty$. If a limit exists, the sequence is said to *converge*; otherwise, it is said to *diverge*.

When the terms of a sequence are added to form a series, then it is often of interest to know whether the sum of the series converges to a limit.

Example 12.1

Let us look at the geometric series $a + ar + ar^2 + ar^3 + \cdots$ In this case define the partial sum S_n by
$$S_n = a + ar + ar^2 + \cdots + ar^n.$$
Therefore,
$$rS_n = ar + ar^2 + \cdots + ar^n + ar^{n+1}.$$
Subtracting the second expression from the first we see that
$$(1-r)S_n = a - ar^{n+1}.$$
Therefore, the partial sum can be written
$$S_n = \frac{a - ar^{n+1}}{(1-r)}, \quad r \neq 1.$$

If $|r| < 1$, then $r^{n+1} \to 0$ as $n \to \infty$, $S_n \to \dfrac{a}{1-r}$ and the series converges.

If $|r| > 1$, then $r^{n+1} \to \infty$ as $n \to \infty$, $S_n \to \infty$, and the series does not converge.

What happens if $r = 1$?

Example 12.2

The divergence of a sequence or series does not necessarily imply that values increase with unbounded magnitude. In Example 12.1, if we take $a = 1$ and $r = -1$, then the values $\{S_n\}$ form the sequence $+1, -1, +1, -1, \ldots$. This behavior is still called divergent, because there is no limit, but it may also be described as *oscillatory*.

Exercise 12.1

Plot the terms in each of the sequences

(a) $\left(\dfrac{1}{2}\right)^n$ (b) $\dfrac{n+1}{n-1}$ (c) $\dfrac{n^2+1}{n^2-1}$

(d) $(2)^n$ (e) $\dfrac{n^2+1}{n-1}$ (f) $\left(1 + \dfrac{1}{n}\right)^n$

Which of the sequences converge?

By evaluating the terms in sequence we can sometimes determine whether it converges. For example, the sequence $\{a_n\}$ for which
$$a_n = 1 + \frac{1}{n}$$
has terms $2, 1.5, 1.333, 1.25, 1.2, \ldots$, which converge to 1. By choosing n sufficiently large, we can make the terms of this sequence as close to 1 as we please. For example, to make the difference between the sequence and the limit 1 less than 0.001, we need to take $n > 1000$, for then
$$|a_n - 1| < 0.001$$
because
$$|a_n - 1| = \frac{1}{n}.$$
To make the difference between the members of the sequence and the limit less than 0.00001, we need to find an n_0 such that
$$|a_n - 1| < 0.00001$$
whenever $n > n_0$. Taking $n_0 = 100000$ will be sufficient. Now suppose we want to make
$$|a_n - 1| < \varepsilon$$
for any choice of ε, can we still find a suitable value for n_0? In this case, since
$$|a_n - 1| = \frac{1}{n}$$
we select $n_0 > \dfrac{1}{\varepsilon}$. As we can always do this, the sequence converges to 1.

Definition 12.1 A sequence $\{a_n\}$ converges to a limit L if for all $\varepsilon > 0$ there exists an n_0 such that
$$|a_n - L| < \varepsilon$$
whenever $n \geq n_0$.

Exercise 12.2

Determine which of the following sequences converge and give the limit where appropriate.

(a) $\left(\dfrac{1}{2}\right)^n$ (b) $\dfrac{n+1}{n-1}$ (c) $\dfrac{n^2+1}{n^2-1}$

(d) $(2)^n$ (e) $\dfrac{n^2+1}{n-1}$ (f) $\left(1+\dfrac{1}{n}\right)^n$

Exercise 12.3

Which of the sequences are increasing/decreasing and bounded?

(a) $\left(\dfrac{1}{2}\right)^n$ (b) $\dfrac{n+1}{n-1}$ (c) $\dfrac{n^2+1}{n^2-1}$

(d) $(2)^n$ (e) $\dfrac{n^2+1}{n-1}$ (f) $\left(1+\dfrac{1}{n}\right)^n$

Although Definition 12.1 gives a formal interpretation for convergence, it is not easy to apply in practice. Numerous results can be used to investigate convergence, but we shall give just one.

Definition 12.2 A sequence $\{a_n\}$ is said to be *increasing (decreasing)* if $a_n > a_m$ ($a_n < a_m$) whenever $n > m$. A sequence is said to be *monotonic* if it is either increasing or decreasing.

It is clear that the Olympic Pole Vault records in Table 12.1 give an increasing sequence. The 1500 meter records given in Table 12.2 are a decreasing sequence.

We now state the following result without proof.

Theorem 12.1 An increasing (decreasing) sequence that is bounded above (below) converges.

Notice, that this result tells us when a sequence converges but does not tell us the limit.

Example 12.3

The record for the 1500 meters is a decreasing sequence that is clearly bounded below by 0. This is a convergent sequence.

Let us now look at the convergence of infinite series. Fortunately, we can use much of what we know about infinite sequences to help us understand series. To see this let us look at the infinite series

$$\sum_{k=1}^{k=\infty} \dfrac{1}{k} = 1 + \dfrac{1}{2} + \dfrac{1}{3} + \dfrac{1}{4} + \cdots$$

The nth partial sum of this series is

$$S_n = \sum_{k=1}^{k=n} a_k,$$

where $a_k = \dfrac{1}{k}$. To demonstrate that the infinite series converges we have to show that the sequence $\{S_n\}$ converges. We have already seen an example of this with the geometric series.

Let's look at some examples. We will concentrate on just two infinite series for the moment:

$$\sum_{k=1}^{k=\infty} \dfrac{1}{k} = 1 + \dfrac{1}{2} + \dfrac{1}{3} + \dfrac{1}{4} + \cdots$$

and

$$\sum_{k=1}^{k=\infty} \dfrac{1}{k^2} = 1 + \dfrac{1}{4} + \dfrac{1}{9} + \dfrac{1}{16} + \cdots$$

Notice that in both cases the individual terms in the series are all positive and tend to zero. Is this sufficient to guarantee that the series converges? The answer, unfortunately, is no.

To see this, look at the first series and rearrange it as follows:

$$\sum_{k=1}^{k=\infty} \frac{1}{k} = 1 + \frac{1}{2} + \left(\frac{1}{3}+\frac{1}{4}\right) + \left(\frac{1}{5}+\frac{1}{6}+\frac{1}{7}+\frac{1}{8}\right) + \cdots$$

$$\geq 1 + \frac{1}{2} + \left(\frac{1}{2}\right) + \left(\frac{1}{2}\right) + \cdots$$

and so we can make the infinite sum larger than any number we please. Since the sequence of partial sums is not bounded above, the corresponding series diverges.

Now consider

$$\sum_{k=1}^{k=\infty} \frac{1}{k^2} = 1 + \frac{1}{4} + \frac{1}{9} + \frac{1}{16} + \cdots.$$

Does the same principle apply here? If we look at the partial sums of this series, it gives an increasing sequence, but we can show that

$$\sum_{k=2}^{k=\infty} \frac{1}{k^2} \leq \int_{1}^{\infty} \frac{1}{x^2} dx = 1$$

so that the partial sums are increasing and bounded above. From this we can conclude that this series converges, though at the moment we cannot state its limit. To get the above upper bound we simply replace the integral by a lower sum using an interval width 1, and notice that the area in each panel is just the corresponding term in the series as shown in Figure 12.2.

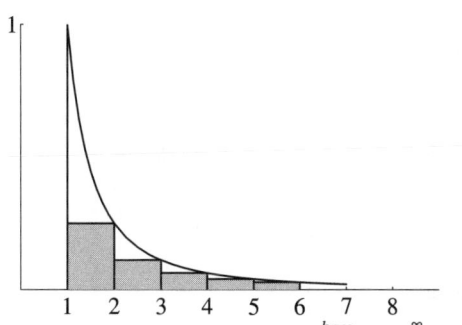

Figure 12.2. Cauchy's Integral Test: $\sum_{k=2}^{k=\infty} \frac{1}{k^2} \leq \int_{1}^{\infty} \frac{1}{x^2} dx$.

This is an example of a useful test for considering convergence.

Theorem 12.2 (Cauchy's Integral Test)
If there exists a function $f(x)$ that is positive, continuous, and decreasing for $x \geq a$ and $f(k) = a_k$ then

$$\sum_{k=1}^{\infty} a_k \quad \text{and} \quad \int_{a}^{\infty} f(x)\, dx$$

either both converge or both diverge.

Special Constants

Many of the most useful constants of mathematics and physics can be expressed as the limits of either a sequence or series, or in some cases both.

Example 12.4
The base of the natural logarithms, e, is given by the limit of the sequence

$$\left(1+\frac{1}{n}\right)^n, \quad n \to \infty.$$

To see this recall that the derivative of the natural log function, $\ln(x)$, is $1/x$ and that when $x=1$ the slope of the curve is 1.

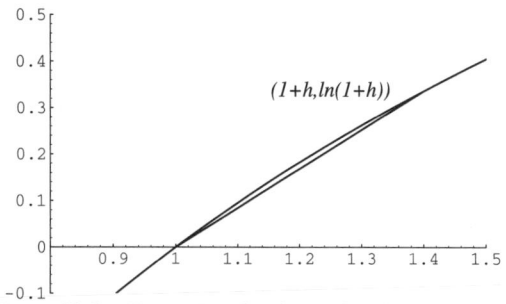

Figure 12.3. Estimating the slope of $y=\ln(x)$ at $x=1$.

The slope of the secant joining $(1, \ln(1))$ to $(1+h, \ln(1+h))$, shown in Figure 12.3, is

$$\frac{\ln(1+h) - \ln(1)}{h} = \ln(1+h)^{\frac{1}{h}}.$$

Now set $n = \frac{1}{h}$, then

$$\underset{h\to 0}{\text{Lim}}\left(\frac{\ln(1+h)-\ln(1)}{h}\right) = \underset{h\to 0}{\text{Lim}}\left(\ln(1+h)^{\frac{1}{h}}\right)$$

$$= \underset{n\to\infty}{\text{Lim}}\left(\ln\left(1+\frac{1}{n}\right)^n\right) = 1.$$

But $\ln(e) = 1$ and so

$$\underset{n\to\infty}{\text{Lim}}\left(\left(1+\frac{1}{n}\right)^n\right) = e.$$

Exercise 12.4

Use the software to show that the sequence $\{a_n\}$ where

$$a_n = \left(1+\frac{1}{n}\right)^n$$

is increasing and bounded above. Estimate e correct to three decimal places. How many terms of this sequence are needed to get e correct to five decimal places?

Example 12.5
We have seen that we need a very large number of terms to get an accurate estimate for e using the sequence in Exercise 12.4. We can also show that[1]

$$e = 1 + 1 + \frac{1}{2!} + \frac{1}{3!} + \frac{1}{4!} + \cdots$$

or

$$e = 1 + \sum_{k=1}^{k=\infty} \frac{1}{k!}.$$

---

[1] The notation $k!$, (k factorial), is shorthand for the product of all the positive integers less than or equal to I. 2!=1.2, 3!=1.2.3, 6!=1.2.3.4.5.6, etc.

Exercise 12.5

Use the software to show that the sequence of partial sums of the series

$$\sum_{k=1}^{k=\infty} \frac{1}{k!}$$

is increasing and bounded above. Estimate e correct to three decimal places. How many terms of the above series are needed to get e correct to five decimal places?

Example 12.6
We have already seen that the series

$$1 + \frac{1}{4} + \frac{1}{9} + \frac{1}{16} + \cdots$$

converges. It is possible to show that the sum of this series is given by

$$\sum_{k=1}^{k=\infty} \frac{1}{k^2} = \frac{\pi^2}{6}$$

which can be used to estimate π.

Exercise 12.6

Use the software to investigate the series

$$\sum_{k=1}^{k=\infty} \frac{1}{k^2}.$$

Use this series to estimate π correct to three decimal places.

Example 12.7
We have previously examined the convergence of sequences for which each term, a_n, is given explicitly. Sometimes, they are given implicitly. For example, let a_0 be any value and define a sequence by

$$a_{n+1} = \frac{1}{2}\left(a_n + \frac{2}{a_n}\right)$$

Taking $a_0 = 1$ gives

$$a_1 = \frac{1}{2}\left(a_0 + \frac{2}{a_0}\right) = \frac{1}{2}(1+2) = 1.5$$

$$a_2 = \frac{1}{2}\left(a_1 + \frac{2}{a_1}\right) = \frac{1}{2}\left(\frac{3}{2} + \frac{4}{3}\right) = 1.41666$$

and so on. This sequence converges to $\sqrt{2}$.

Exercise 12.7

Use the above sequence to estimate $\sqrt{2}$ correct to five decimal places. How many additional iterations are needed to get nine decimal places correct?

Series Expansions

We have already seen that the geometric series
$$1 + r + r^2 + r^3 + \cdots$$
converges to
$$\frac{1}{1-r}$$
provided $|r|<1$. Therefore, we can write the rational function $\frac{1}{1-x}$ as a series in terms of x, that is,
$$\frac{1}{1-x} = 1 + x + x^2 + \cdots, \quad |x|<1.$$

This is called a *power series expansion*. We will now show that most smooth functions can be written in this way.

Let $f(x)$ be a smooth function and assume that we can write
$$f(x) = a_0 + a_1 x + a_2 x^2 + \cdots$$

Taking $x=0$ gives
$$f(0) = a_0.$$

Next we differentiate the power series with respect to x to get
$$f'(x) = a_1 + 2a_2 x + \cdots$$

(It is possible to show that, provided the series converges, we can differentiate an infinite series term by term.) Now set $x=0$ in this series to give
$$f'(0) = a_1.$$

We repeat this again and again to get
$$f''(0) = 2 \times a_2 = 2!\, a_2,$$
$$f'''(0) = 3 \times 2 \times a_3 = 3!\, a_3.$$

Combining these results gives
$$f(x) = f(0) + xf'(0) + \frac{1}{2!}x^2 f''(0)$$
$$+ \frac{1}{3!}x^3 f'''(0) + \cdots$$

This is called the Maclaurin expansion of the function $f(x)$. It tells us that if we knew all the derivatives of a function at $x=0$, we could write down the value of the function at any other point from this infinite series. Of course, it is necessary to show that the series converges before we can do this.

Example 12.8

The derivatives of $f(x) = \dfrac{1}{1-x}$ at $x=0$ are
$$f'(0) = 1,$$
$$f''(0) = 2,$$
$$f'''(0) = 3!,$$
$$f^{(iv)}(0) = 4!,$$

so that the Maclaurin expansion of f is
$$f(x) = 1 + x + x^2 + x^3 + x^4 + \cdots$$

This is just the standard geometric series, and we know that this series only converges when $|x|<1$.

The Maclaurin expansion is a particular example of a more general result that produces the following expansion:

$$f(x) = f(a) + (x-a)f'(a) + \frac{1}{2!}(x-a)^2 f''(a) + ..$$

This is called a Taylor series sxpansion. If we set $a=0$, we recover the Maclaurin expansion.

To see where this series comes from, recall that the derivative of a function $f(x)$ at a point $x=a$ is defined by

$$f'(a) = \lim_{h \to 0} \left(\frac{f(a+h) - f(a)}{h} \right).$$

Now set $x=a+h$; then this becomes

$$f'(a) = \lim_{x \to a} \left(\frac{f(x) - f(a)}{x - a} \right).$$

Let us now suppose that we do not take the limit but instead write

$$f'(a) = \left(\frac{f(x) - f(a)}{x - a} \right) + E$$

where E tends to zero as $x \to a$. Rearranging this expression gives

$$f(x) = f(a) + (x-a)f'(a) + (x-a)E,$$

which gives the first two terms of the Taylor series expansion of f about $x=a$. If a and x are sufficiently close that E is small, then we have a *linear approximation* for $f(x)$ near $x=a$, that is

$$f(x) \approx f(a) + (x-a)f'(a).$$

We could also obtain the Taylor series by looking for an expansion of the form

$$f(x) = a_0 + (x-a)a_1 + (x-a)^2 a_2 + \cdots$$

Setting $x=a$ gives $a_0 = f(a)$. Next we differentiate and put $x=a$ to give $a_1 = f'(a)$. Repeated differentiation gives the Taylor series expansion.

Transcendental Expansions

We have already seen, in Example 12.8, how the Maclaurin series expansion of the function $f(x)=1/(1-x)$ gave the geometric series. Now we will see how Taylor series can be used to obtain power series expansions for some of the usual transcendental functions.

Example 12.9

The power series expansion of e^x about $x=0$ is given as follows. Since the derivative of e^x is e^x, then $f(0) = f'(0) = f''(0) = ... = 1$ and so

$$f(x) = f(0) + xf'(0) + \frac{1}{2!}x^2 f''(0)$$
$$+ \frac{1}{3!}x^3 f'''(0) + \cdots$$

becomes

$$e^x = 1 + x + \frac{1}{2!}x^2 + \frac{1}{3!}x^3 + \cdots$$

In particular

$$e = 1 + 1 + \frac{1}{2!} + \frac{1}{3!} + \cdots$$

which is the series we used in Exercise 12.5 to evaluate e. It can be shown that this series converges for all values of x, but that many terms will be needed if x is much bigger than 1.

Example 12.10

The power series expansion (Maclaurin expansion), of $\sin(x)$ about $x=0$ is given as follows. First, $f(0)=\sin(0)=0$. Then, since the derivative of $f(x) = \sin(x)$ is $\cos(x)$, we have that $f'(0) = 1$. Similarly, $f''(x) = -\sin(x)$ gives $f''(0) = 0$, and so on to produce

$$\sin(x) = x - \frac{1}{3!}x^3 + \frac{1}{5!}x^5 - \cdots$$

Remember that x is measured in radians. This series converges for all values of x.

Example 12.11

The power series expansion, Taylor expansion, of $f(x)=\cos(x)$ about $x=\frac{\pi}{2}$ is given as follows. First, $f\left(\frac{\pi}{2}\right) = \cos\left(\frac{\pi}{2}\right) = 0$. Next, since the derivative of $f(x) = \cos(x)$ is $-\sin(x)$, we get $f'\left(\frac{\pi}{2}\right) = -1$.

Similarly, $f''(x) = -\cos(x)$, hence, $f''\left(\frac{\pi}{2}\right) = 0$, and so on to give

$$\cos(x) = -\left(x - \frac{\pi}{2}\right) + \frac{1}{3!}\left(x - \frac{\pi}{2}\right)^3 - \frac{1}{5!}\left(x - \frac{\pi}{2}\right)^5 + \cdots$$

Remember that x is measured in radians.

Exercise 12.8

Find Maclaurin expansions of the following functions:
   (a) $\cos(x)$,
   (b) $\ln(1+x)$.

Exercise 12.9

Explain why it is not possible to find a Maclaurin expansion of $\ln(x)$.

Exercise 12.10

Find Taylor series expansions of the following functions about the points given:
   (a) $\sin(x)$, $x=0.2$,
   (b) $\ln(1+x)$, $x=1$.
Estimate the values of x for which these series converge.

Shortcuts to Taylor Series

We have outlined two ways to obtain a Taylor series expansion. The first involves the general expression

$$f(x) = f(a) + (x-a)f'(a) + \frac{1}{2!}(x-a)^2 f''(a) + \cdots$$

and can be found by evaluating the derivatives of f at $x=a$. Alternatively, we can start with the general series

$$f(x) = a_0 + a_1(x-a) + a_2(x-a)^2 + \cdots$$

and by successively differentiating we can find the coefficients. Both methods can be very time consuming. We will now outline two possible shortcuts.

Example 12.12

The Taylor expansion of e^x about $x=0$ is given by

$$e^x = 1 + x + \frac{x^2}{2!} + \frac{x^3}{3!} + \cdots$$

We can adapt this series to find the Taylor expansions of e^{-x} and e^{-x^2} as follows

$$e^{-x} = 1 + (-x) + \frac{(-x)^2}{2!} + \frac{(-x)^3}{3!} + \cdots$$

$$= 1 - x + \frac{x^2}{2!} - \frac{x^3}{3!} + \cdots$$

$$e^{x^2} = 1 + (x^2) + \frac{(x^2)^2}{2!} + \frac{(x^2)^3}{3!} + \cdots$$

$$= 1 + x^2 + \frac{x^4}{2!} + \frac{x^6}{3!} + \cdots$$

Definition 12.3 A function, $f(x)$, is said to be
$$\text{\emph{even} if } f(-x) = f(x)$$
and
$$\text{\emph{odd} if } f(-x) = -f(x).$$

Example 12.13
(a) The function $\cos(x)$ is an even function, and $\sin(x)$ is an odd function.

(b) The function e^x is neither odd nor even.

We can use even and oddness to simplify the procedure of obtaining a series expansion about $x=0$ since the series expansion of an even function will contain only even powers and an odd function only odd powers.

Example 12.14
The function $f(x) = e^{x^2}$ is an even function since

$$f(-x) = e^{(-x)^2} = e^{x^2} = f(x)$$

and so its series expansion involves only even powers, that is,

$$e^{x^2} = a_0 + a_2 x^2 + a_4 x^4 + \cdots.$$

Setting $x=0$ gives $a_0 = 1$. Differentiating twice and setting $x=0$ gives $a_2 = 1$. Differentiating twice again and setting $x=0$ gives $a_4 = \frac{1}{2!}$. Repeating this process gives the series expansion given in Example 12.12.

Applications of Expansions

Every quadratic equation has at most two solutions, each cubic has at most three, and each quartic has at most four. Just as it is possible to find a formula to solve quadratics, there is a formula to solve any cubic or quartic equation. However, the equation

$$\cos(x) = x$$

is called a transcendental equation, and there are very few results that help us to solve such equations analytically. If we plot the functions $y=\cos(x)$ and $y=x$, then the curves meet at only one point near $x=0.6$ where $\cos(x)=x$. (See Figure 12.4.) How can we find this value?

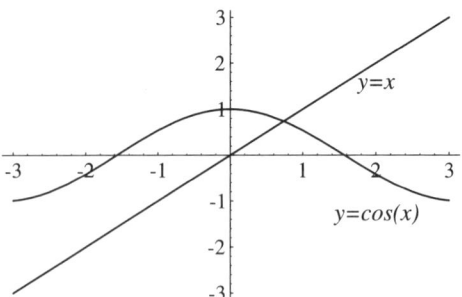

Figure 12.4. The solution of the equation $\cos(x)=x$.

If we expand $\cos(x)$ in a Maclaurin series about $x=0$, then $\cos(x) \approx 1 - \frac{1}{2}x^2$, and we can approximate the original equation by the quadratic equation

$$x = 1 - \frac{1}{2}x^2$$

or

$$x^2 + 2x - 2 = 0.$$

This quadratic equation has the solutions $x=0.732051$ and $x=-2.73205$. The first is clearly the solution we seek; but where does the second solution come from?

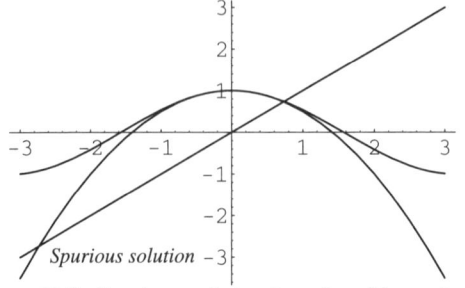

Figure 12.5. Spurious solution introduced by replacing $\cos(x)$ by $1-\frac{x^2}{2}$.

If we plot the curve $y = 1 - \frac{x^2}{2}$ along with $\cos(x)$ and x, we see that introducing the approximation has given a second point where the curves meet. Clearly, this is a spurious solution that is not relevant to the original equation. (See Figure 12.5.)

The Binomial Theorem

The Binomial Theorem helps us to expand expressions of the form $(a+x)^n$. Recall that

$$(a+x)^2 = a^2 + 2ax + x^2$$
$$(a+x)^3 = a^3 + 3a^2x + 3ax^2 + x^3$$
$$(a+x)^4 = a^4 + 4a^3x + 6a^2x^2 + 4ax^3 + x^4$$

and in general

$$(a+x)^n = a^n + na^{n-1}x + \tfrac{n(n-1)}{2}a^{n-2}x^2 + \cdots$$
$$+ \tfrac{n(n-1)}{2}a^2x^{n-2} + nax^{n-1} + a^n.$$

This is called the *Binomial* expansion of $(a+x)^n$. We can obtain this expression as the Maclaurin expansion of $(a+x)^n$.

Now let us consider the function
$$f(x) = (a+x)^n,$$
for which $f(0) = a^n$. Differentiating f gives
$$f'(x) = n(a+x)^{n-1},$$
and so
$$f'(0) = na^{n-1}.$$
Similarly,

$$f''(x) = n(n-1)(a+x)^{n-2},$$
$$f''(0) = n(n-1)a^{n-2},$$
$$\vdots$$
$$f^{(j)}(x) = n(n-1)\cdots(n-j+1)(a+x)^{n-j}, j \leq n,$$
$$f^{(j)}(0) = n(n-1)\cdots(n-j+1)a^{n-j}, j \leq n,$$
$$f^{(n+1)}(0) = 0, f^{(n+2)}(0) = 0, \cdots$$

Substituting into the general Taylor series gives

$$f(x) = f(0) + xf'(0) + \tfrac{x^2}{2}f''(0) + \cdots$$
$$+ \tfrac{x^n}{n!}f^{(n)}(0) + \tfrac{x^{n+1}}{(n+1)!}f^{(n+1)}(0) + \cdots$$

which simplifies to

$$f(x) = a^n + na^{n-1}x + \tfrac{n(n-1)}{2}a^{n-2}x^2 + \cdots + x^n$$

and we have proven the Binomial Theorem for positive integer values of n.

The Maclaurin also helps us to expand expressions of the form $(a+x)^n$ when n is not a positive integer or when it is not an integer at all.

Example 12.15

Find a Maclaurin series expansion for $\sqrt{(1+x)}$ and use the result to estimate $\sqrt{150}$.

Define $f(x) = \sqrt{(1+x)}$, then $f(0) = 1$. Similarly,

$$f'(x) = \tfrac{1}{2}\tfrac{1}{\sqrt{(1+x)}}, \quad f'(0) = \tfrac{1}{2},$$

$$f''(x) = -\tfrac{1}{4}\tfrac{1}{\sqrt{(1+x)^3}}, \quad f''(0) = -\tfrac{1}{4},$$

$$f'''(x) = \tfrac{3}{8}\tfrac{1}{\sqrt{(1+x)^5}}, \quad f'''(0) = \tfrac{3}{8},$$

and so on. Combining these results gives

$$f(x) = 1 + \tfrac{1}{2}x - \tfrac{1}{8}x^2 + \tfrac{1}{16}x^3 - \tfrac{5}{128}x^4 + \cdots.$$

We can use this result to evaluate $\sqrt{150}$ as follows. If we write $\sqrt{150}$ as

$$\sqrt{150} = \sqrt{144+6} = \sqrt{144\left(1+\tfrac{6}{144}\right)}$$
$$= 12\sqrt{1+\tfrac{1}{24}} = 12 f(\tfrac{1}{24})$$

and then take the first four terms in the series expansion for f, we obtain

$$12 f(\tfrac{1}{24}) \approx 12.24745009$$

compared with $\sqrt{150} \approx 12.24744871$.

WORKED EXAMPLE 12.1
SEQUENCES

Many definite integrals cannot be evaluated since no indefinite integral can be found in terms of elementary functions. We have already seen one technique for approximating such integrals, "Simpson's Rule", in Module (11). We can also use series expansions.

Let us consider the integral

$$\int_0^1 e^{-x^2} dx.$$

We can estimate the values of this integral by approximating the integrand e^{-x^2} by its Maclaurin expansion, as shown in Figure 12.6, and integrating the polynomial which results.

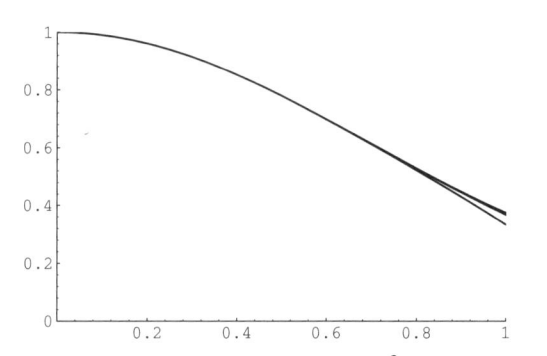

Figure 12.6. The function $f(x) = e^{-x^2}$ and its Maclaurin expansion of order 6.

To find the Maclaurin expansion of $f(x) = e^{-x^2}$ we could evaluate the derivatives of f at $x=0$ and construct the series from the general expression. Alternatively, the power series expansion for e^x is

$$e^x = 1 + x + \frac{1}{2!}x^2 + \frac{1}{3!}x^3 + \cdots$$

and replacing x by $-x^2$ gives

$$e^{-x^2} = 1 - x^2 + \frac{1}{2}x^4 - \frac{1}{6}x^6 + \frac{1}{24}x^8 - \cdots$$

Let us approximate e^{-x^2} by the first four terms of this series, that is,

$$e^{-x^2} \approx 1 - x^2 + \frac{1}{2}x^4 - \frac{1}{6}x^6$$

then

$$\int_0^1 e^{-x^2} dx \approx \int_0^1 \left(1 - x^2 + \frac{1}{2}x^4 - \frac{1}{6}x^6\right) dx$$

$$= \left[x - \frac{x^3}{3} + \frac{x^5}{10} - \frac{x^7}{42}\right]_0^1$$

$$= 0.742857.$$

Including another term gives

$$\int_0^1 e^{-x^2} dx \approx \int_0^1 \left(1 - x^2 + \frac{1}{2}x^4 - \frac{1}{6}x^6 + \frac{1}{24}x^8\right) dx$$

$$= \left[x - \frac{x^3}{3} + \frac{x^5}{10} - \frac{x^7}{42} + \frac{x^9}{216}\right]_0^1$$

$$= 0.747487.$$

The value of the integral is 0.746824, to six decimal places, so we are still some way from the exact result. It takes a series involving x^{16} to get six decimal places correct, nevertheless, this form of integration can be very useful.

WORKED EXAMPLE 12.2
NEWTON'S METHOD

Suppose we wish to solve the equation

$$x^3 - 3x^2 + 4 = 0.$$

This is a cubic equation and has at most three solutions. We know how to deal with quadratic equations; there is a *simple* formula. A formula for finding the solutions of any cubic equation was discovered in the 17th century by Tartaglia and published amid great controversy by Cardano. However, their formula is difficult to use. We will examine an alternative method attributed to Sir Isaac Newton.

In general, we are looking for a solution of the equation $f(x)=0$ at a point $x=a$. Suppose we have an approximate solution that we write as x_0, then we can write $a = x_0 + \varepsilon$. Since $f(a)=0$; we have that

$$0 = f(a) = f(x_0 + \varepsilon).$$

Expanding $f(x_0 + \varepsilon)$ in a Taylor series about $x = x_0$ gives

$$0 = f(a) = f(x_0 + \varepsilon)$$
$$= f(x_0) + \varepsilon f'(x_0) + \frac{\varepsilon^2}{2} f''(x_0) + \cdots$$

Whenever ε is small, we can ignore the terms beyond $\varepsilon f'(x_0)$ and solve for ε to get

$$\varepsilon \approx -\frac{f(x_0)}{f'(x_0)},$$

provided $f'(x_0) \neq 0$. Therefore,

$$a \approx x_0 - \frac{f(x_0)}{f'(x_0)}.$$

If we write $x_1 = x_0 - \dfrac{f(x_0)}{f'(x_0)}$, then x_1 should be a better approximation for the solution at $x=a$. We can then repeat this process.

This defines a sequence of approximations

$$x_{n+1} = x_n - \frac{f(x_n)}{f'(x_n)}.$$

This is called *Newton's method*. If the sequence of values converges then it tends to the solution of the equation $f(x)=0$.

Setting
$$f(x) = x^3 - 3x^2 + 4$$

then
$$f(-2) = -16 < 0, \; f(0) = 4 > 0$$

and so there must be a value in the interval $[-2, 0]$ where f is zero and the equation $f(x)=0$ is satisfied.

If we take as an initial guess, $x_0 = -0.9$, then Newton's method generates the sequence

$$x_1 = x_0 - \frac{f(x_0)}{f'(x_0)} = -0.9 - \frac{0.841}{7.83} = -1.00741,$$

$$x_2 = x_1 - \frac{f(x_1)}{f'(x_1)} = -1.00004,$$

$$x_3 = x_2 - \frac{f(x_2)}{f'(x_2)} = -1.00000000.$$

Any subsequent iterations give the same result. This sequence has converged in just three iterations.

We can use Taylor series to explain why this rapid convergence is attained. Let us suppose that x_n is an estimate for the solution of $f(x) = 0$ at $x = a$; then we can write

$x_n = a + e_n$, where e_n is the error in this estimation. Now set $x_{n+1} = a + e_{n+1}$ and substitute into Newton's method to get

$$a + e_{n+1} = x_n - \frac{f(x_n)}{f'(x_n)}$$

$$= a + e_n - \frac{f(a+e_n)}{f'(a+e_n)}$$

or

$$e_{n+1} = e_n - \frac{f(a+e_n)}{f'(a+e_n)}.$$

Next, expand $f(a+e_n)$ and $f'(a+e_n)$ as Taylor series to give

$$f(a+e_n) = f(a) + e_n f'(a) + \tfrac{1}{2} e_n^2 f''(a) + \cdots$$

and

$$f'(a+e_n) = f'(a) + e_n f''(a) + \tfrac{1}{2} e_n^2 f'''(a) + \cdots$$

then

$$e_{n+1} = e_n - \frac{f(a) + e_n f'(a) + \tfrac{1}{2} e_n^2 f''(a) + \cdots}{f'(a) + e_n f''(a) + \tfrac{1}{2} e_n^2 f'''(a) + \cdots}$$

$$= \frac{\tfrac{1}{2} e_n^2 f''(a) + \cdots}{f'(a) + e_n f''(a) + \tfrac{1}{2} e_n^2 f'''(a) + \cdots}.$$

Finally, provided $f'(a) \neq 0$ and e_n is small, we get

$$e_{n+1} \approx \tfrac{1}{2} e_n^2 \frac{f''(a)}{f'(a)}.$$

This is a significant result because it tells us that after each iteration with Newton's method the error is squared. For example, if the error at any stage is 0.01, then the error after the next iteration will be approximately $0.01^2 = 0.0001$, and 0.00000001 at the next and so on. This explains the very rapid convergence of Newton's method for this example.

PROBLEM 12.1
SEQUENCES AND
SERIES

Name: _____
Date: _____
Section: _____

1. Investigate the following sequences and tick the appropriate boxes.

	Increasing	Decreasing	Bounded Above	Bounded Below	Convergent	Limit
(a) $\dfrac{n+1}{n^2-2}$	[ ]	[ ]	[ ]	[ ]	[ ]	
(b) $\dfrac{n^2+1}{n-2}$	[ ]	[ ]	[ ]	[ ]	[ ]	
(c) $\dfrac{(-1)^n}{n-2}$	[ ]	[ ]	[ ]	[ ]	[ ]	
(d) $\dfrac{(n)^n}{n-2}$	[ ]	[ ]	[ ]	[ ]	[ ]	
(e) $\dfrac{(2)^n}{n^2}$	[ ]	[ ]	[ ]	[ ]	[ ]	

2. Which of the following series converge and why?

(a) $\displaystyle\sum_{k=1}^{k=\infty} \dfrac{\left(\frac{1}{2}\right)^k}{k^2}$

(b) $\displaystyle\sum_{k=1}^{k=\infty} \dfrac{(-1)^k}{k^2}$

(c) $\displaystyle\sum_{k=1}^{k=\infty} \dfrac{1}{\sqrt{k}}$

66 Calculus Connections

3. Determine the power series expansions of the following functions about the points given. In each case, estimate the values of x for which the expansion is valid.

(a) $\tan(x)$, $x=0$,

(b) $\ln(1-x)$, $x=\frac{1}{2}$,

(c) $\sin(x)$, $x=\pi$,

(d) $\ln(x)$, $x=1$.

4. Use the Maclaurin and Taylor expansions previously developed to find expansions for

(a) $\sin(x^2)$, $x=0$,

(b) $\ln(1-t^3)$ $t=0$,

(c) $\cos(\sqrt{x})$, $x=0$,

(d) $\ln(\frac{1-x}{1+x})$, $x=\frac{1}{2}$.

PROBLEM 12.2
NEWTON'S METHOD

Name: _____

Date: _____

Section: _____

In Worked Example 12.2 we looked at Newton's method for solving an equation of the form $f(x) = 0$. If x_n is an estimate of a solution at $x = a$, we saw that the sequence of approximations generated by

$$x_{n+1} = x_n - \frac{f(x_n)}{f'(x_n)}$$

converged very rapidly. We also saw the reason for this, but the mathematics depended on the condition that $f'(a) \neq 0$. In this problem we will explore what this condition means.

1. Use Newton's method to solve the equation $x^3 - 3x^2 + 4 = 0$, taking an initial guess of $x_0 = 1.9$.

$$f'(x) = \quad [\qquad\qquad]$$
$$f(x_0) = f(1.9) = [\qquad\qquad]$$
$$f'(x_0) = f'(1.9) = [\qquad\qquad]$$
$$x_1 = \quad [\qquad\qquad]$$
$$x_2 = \quad [\qquad\qquad]$$
$$\vdots$$
$$x_{10} = \quad [\qquad\qquad]$$

In Worked Example 12.2, Newton's method converged in 3 iterations. Do we get the same rapid convergence here? If not, why not?

2. Plot the function $f(x) = x^3 - 3x^2 + 4$ for $-2 \leq x \leq 3$. What do you notice about the derivative of f near $x \approx 1.9$?

3. From Worked Example 12.2 we saw that

$$e_{n+1} = e_n - \frac{f(a)+e_n f'(a)+\frac{1}{2}e_n^2 f''(a)+\cdots}{f'(a)+e_n f''(a)+\frac{1}{2}e_n^2 f'''(a)+\cdots}$$

$$= \frac{\frac{1}{2}e_n^2 f''(a)+\cdots}{f'(a)+e_n f''(a)+\frac{1}{2}e_n^2 f'''(a)+\cdots}$$

which reduced to

$$e_{n+1} \approx \frac{1}{2}e_n^2 \frac{f''(a)}{f'(a)}$$

provided $f'(a) \neq 0$. If $f'(a) = 0$, show that

$$e_{n+1} \approx \frac{1}{2}e_n.$$

How does this explain the behavior of Newton's method in this case?

NEW SITUATIONS

1. Square and cube roots

In Example 12.7 we saw that the sequence

$$a_{n+1} = \frac{1}{2}\left(a_n + \frac{2}{a_n}\right)$$

converged rapidly to $\sqrt{2}$. Show that this is equivalent to applying Newton's method to solve the equation

$$x^2 - 2 = 0.$$

Extend this method to determine the square root of any number, b, by solving the equation

$$x^2 - b = 0.$$

How would you use Newton's method to find cube and higher roots?

2. Finding limits

In Chapter 2 we looked at the limits of functions in order to determine more about their behavior. For example, the behavior of the function

$$\frac{\sin(x)}{x}$$

as x tends to 0 is unclear. We can use series expansions to investigate this. The Maclaurin expansion of $\sin(x)$ is

$$\sin(x) = x - \frac{1}{3!}x^3 + \frac{1}{5!}x^5 - \cdots$$

so that

$$\frac{\sin(x)}{x} = \frac{x - \frac{1}{3!}x^3 + \frac{1}{5!}x^5 - \cdots}{x}$$

$$= 1 - \frac{1}{3!}x^2 + \frac{1}{5!}x^4 - \cdots$$

and now it is clear that

$$\frac{\sin(x)}{x} \to 1, \quad x \to 0.$$

Investigate the following:

(a) $e^{\frac{1}{x}}$, $x \to 0$,

(b) $\cos(\sqrt{x})$, $x \to 0$,

(c) $\frac{\cos(x)}{2x - \pi}$, $x \to \frac{\pi}{2}$.

3. L'Hôpital's rule

In the previous exercise we looked at the function

$$\frac{\sin(x)}{x}$$

and used Taylor series expansions to show that

$$\frac{\sin(x)}{x} \to 1, \quad x \to 0.$$

We could have obtained this from a more general result known as L'Hôpital's Rule which states that if f and g are differentiable functions such that

$$\mathop{\mathrm{Lim}}_{x \to a}(f(x)) = 0$$

and

$$\mathop{\mathrm{Lim}}_{x \to a}(g(x)) = 0$$

then

$$\mathop{\mathrm{Lim}}_{x \to a}\left(\frac{f(x)}{g(x)}\right) = \mathop{\mathrm{Lim}}_{x \to a}\left(\frac{f'(x)}{g'(x)}\right)$$

For
$$\frac{\sin(x)}{x}$$
$$\lim_{x \to 0}\left(\frac{\sin(x)}{x}\right) = \lim_{x \to 0}\left(\frac{\cos(x)}{1}\right) = 1.$$

Use Taylor series expansions of f and g to justify this result.

Historical Note: L'Hôpital's Rule was actually discovered by John Bernoulli but given to Guillaume Francois Antoine de L'Hôpital (Marquis de St. Mesme) in return for salaried employment!

4. Remainder form of Taylor's expansion

We have seen that for many functions we can determine a Taylor series expansion

$$f(x) = f(a) + (x-a)f'(a) + \frac{1}{2}(x-a)^2 f''(a) + \cdots$$

However, this is an infinite series and can be difficult to manipulate. An alternative form is

$$f(x) = f(a) + (x-a)f'(a) + \frac{1}{2!}(x-a)^2 f''(a)$$
$$+ \cdots + \frac{1}{n-1!}(x-a)^{n-1} f^{(n-1)}(a) + R_n$$

which is called the *remainder form* of the Taylor series. We can show that

$$R_n = \frac{1}{n!}(x-a)^n f^{(n)}(\xi)$$

where ξ is an unknown value between x and a. To derive this result, we write

$$\int_a^x f'(x)dx = f(x) - f(a)$$

or

$$f(x) = f(a) + \int_a^x f'(x)dx$$

Now apply the Mean Value Theorem from Chapter 8 to give

$$f(x) = f(a) + (x-a)f'(\xi)$$

which is the remainder form of the Taylor series with $n=1$. Similarly,

$$\int_a^x f''(x)dx = f'(x) - f'(a)$$

so that

$$f'(x) = f'(a) + \int_a^x f''(x)dx$$
$$= f'(a) + (x-a)f''(\xi).$$

Integrating both sides with respect to x gives

$$f(x) = f(a) + (x-a)f'(a) + \frac{1}{2}(x-a)^2 f''(\xi).$$

By evaluating the integral $\int_a^x f^{(n)}(x)dx$, derive the general form of the Taylor series with remainder.

Estimate how many terms will be needed in the Taylor series of e^x about $x=0$ to obtain an approximation that is correct to five decimal places over the interval $|x|<0.1$.

Chapter 13
Differential Equations

Prerequisites

Before you study this material, you should be familiar with:

(1) The elements of differential and integral calculus (Modules 2, 3, 5, 6, and 7).
(2) The concept of definite integration and its applications (Module 9).
(3) Some examples of relations between quantities that involve differentials (Modules 9, 10).

Objectives

Differential equations, and their solution, are foremost examples of the usefulness and relevance of calculus. Many of the topics of the preceding chapters lead up to this discussion: functions, limits, differentiation, and integration. Differential equations have appeared in earlier chapters, such as 9, Definite Integrals and 10, Rectilinear Motion, although they were not termed differential equations at the time.

This chapter explains and illustrates the concepts of direction fields, isoclines, and phase plane plots. We explore the meaning of a differential equation, the order of a differential equation, and the meaning and significance of initial conditions. There are many techniques for solving special forms of differential equations, but only two are given here as examples: separation of variables, and the standard method for homogeneous linear second-order differential equations with constant coefficients. Standard textbooks give details of other techniques.

Connections

The first application illustrated in the software is population modeling.

Population growth is one of the major issues facing the world today. Why are people worried? Quite simply, it is because if the present birth rate continues, the population will grow exponentially until it becomes limited by lack of resources. In an attempt to understand this and similar problems, biologists have studied simple organisms like bacteria and fruit flies, as well as human populations. We will start with the simplest model, which assumes a population increasing at a constant rate.

At first sight you may not think this is a suitable topic in which to apply calculus, for populations consist of individuals and we would not expect to be able to describe them with continuous variables. Also, they are altered by discrete events (births and deaths), which again would not seem to be appropriately described by continuous variables such as we find used in calculus. However, for large populations the errors in making this approximation should be small. (Even in a small country with a population of, say, near 1 million, the quantitative error in describing the population as 1000000.4 is self-evidently very small.) The error that arises in this way is much smaller than the error that will be introduced by other assumptions, such as supposing that birth and death rates are constant over time. The point is, that despite all the failures of simple models to be totally realistic, it is still possible to learn something useful from the mathematics. We return to mathematical modeling in a later chapter, where we say more about its rationale.

The second application illustrated in the software is a bungee jump. In *Calculus Connections* we have already seen that the movement of various objects can be described mathematically, and the basic principles will be familiar. This application demonstrates how laws of physics (Newton's Second Law of Motion in this case) lead to a differential equation.

We will see that a differential equation alone is never a complete statement of a problem. Differential equations predict how values change, and they therefore require some additional information. In this case we will need to know the initial position and velocity of the jumper. We refer to these values as *initial conditions*.

Introduction to ODEs

Many laws of nature, it seems, are written in terms of rates of change. For example, Newton's Laws include the famous "Force = rate of change of momentum". This is written down mathematically as a differential equation. One of Newton's great triumphs was to show that when the force is given by Newton's Law of Gravity (the inverse square law), then the resulting differential equations predict the motion of the planets. Today, we can add many more examples to the list of the uses of differential equations: anything to do with the mechanics of motion, including space flight; chemical reaction kinetics; biological population modeling; electrical and electronic circuits; even financial markets and the flow of traffic on freeways. In all these cases the mathematical expression of the scientific law links a rate of change to some function of the quantities involved. By solving the equations, a quantitative prediction of the behavior of the system can be obtained.

There is such a huge variety of DEs that we cannot hope in this chapter to do more than introduce the topic for the simplest cases, namely, ordinary differential equations of low order. Ordinary differential equations (ODEs) are differential equations with only one independent variable. Newton's Laws for a planet are an example, where time is the independent variable. "Order" refers to the highest derivative involved, so Newton's Law is second-order, because acceleration is the second derivative of the function $x(t)$. More elaborate differential equations involve more than one independent variable; these are called *partial differential equations* (PDEs). An interesting, if advanced, example is Einstein's Field Equations with four independent variables, three in space plus time; these in principle describe the large-scale structure of our universe and can be paraphrased as "Matter tells space how to bend, space tells matter how to move". A typical example of their importance is the claim by many experts that the PDEs which model a nuclear explosion can be solved so reliably using supercomputers that actual nuclear tests are no longer necessary. In more down-to-earth examples, aerodynamic flow is described by PDEs and modern designers of cars, planes, ships, and so on use supercomputers to solve PDEs and do simulation tests of their designs before starting the very expensive step of building prototypes. PDEs describe the weather, ocean currents, the ozone layers in the stratosphere, and so on.

The immediate objective of this chapter is to look at simple ODEs. First, we shall try to understand what a differential equation is telling us, which we will do by thinking of it as a rule for taking a small step at a time. Often the interpretation of these steps is closely linked to the science involved. We will look at some general methods for trying to understand qualitatively how the solution of an ODE behaves, such as direction fields, and see why initial conditions are needed to find a particular solution.

The objective of this chapter is not to cover systematically all the special techniques for solving ODEs; there are entire textbooks devoted to this. We shall cover only two basic techniques to demonstrate that analytic (closed-form) solution of ODEs is possible in certain cases, and note that analytic solution of a general ODE is not usually possible. All the practical applications mentioned above rely on numerical solution. We will not study the numerical methods needed, but will hint that they are based on the step-by-step approach.

Population Growth: Simple Model

Suppose there is a population (of bacteria, for example) where the birth rate and death rate per member are known. If the population is N, then the population increases by rN per unit of time, where the net rate of increase is r per member.

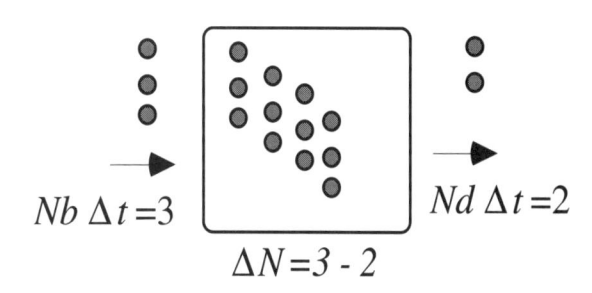

Figure 13.1. Population change.

The diagram in Figure 13.1 shows what these assumptions mean if we can apply them to an interval of time Δt. Let the birth rate per member be b and the death rate be d. If the population has size $N(t)$ at time t, then in the interval Δt, $Nb\Delta t$ individuals will be added to the population (by birth) and $Nd\Delta t$ individuals will leave (by death). This applies approximately if the number N is large so that we can treat N as a continuous quantity.

The population will therefore increase by $\Delta N = (Nb\Delta t - Nd\Delta t) = (b-d)N\Delta t$.

Now we let $r=(b-d)$ and write $\dfrac{\Delta N}{\Delta t} = rN$.

Letting $\Delta t \to 0$, we see that in the limit under these assumptions the population is modeled by

$$\frac{dN}{dt} = rN \qquad (13.1)$$

with $N = N_0$ at $t=0$ as the initial condition.

This model is a simple differential equation, and we will see that its solution is $N = N_0 \exp(rt)$. Over periods of time in which our assumptions remain true this is quite a good model of population growth.

First note that first order differential equations can be put in a standard form. We will use y for the dependent variable and x for the independent variable. The simple population growth model would then be written as

$$\frac{dy}{dx} = ry \quad (13.8)$$

The general form of a first order DE is

$$\frac{dy}{dx} = f(x, y) \quad (13.9)$$

Here are some examples to show what this form means.

Example 13.1

In equation (13.9) take $f(x, y) = ry$. This gives (13.8). In this example the function $f(x,y)$ does not involve x. $y(x) = A\exp(rx)$ is a solution of this equation because

$$y' = Ar\exp(rx)$$

which equals $ry(x)$.

In equation (13.9) take $f(x, y) = \cos(x)$. Now the function $f(x,y)$ does not involve y. $y(x) = A + \sin(x)$ is a solution of this equation because $y' = \cos(x)$.

As an example of a function $f(x,y)$ which involves x and y, take $f(x, y) = x^2 + y^2$. There is no closed-form solution in this case.

Exercise 13.3

Initial condition

Notice that both closed-form solutions given in Example 13.1 have an arbitrary constant in them. This is equivalent to saying that we can start the solution at any point we want. Try this out. By starting at different points, you should be able to generate a whole family of solutions.

To help visualize the solution in advance of the numerical or analytical solution of a differential equation, note that the function $f(x,y)$ specifies the slope of the solution at any point (x,y). Then plot little segments with slope $f(x,y)$ at a grid of points as in Figure 13.3. The resulting diagram is called a *direction field*. The possible shapes of the solution will suggest themselves.

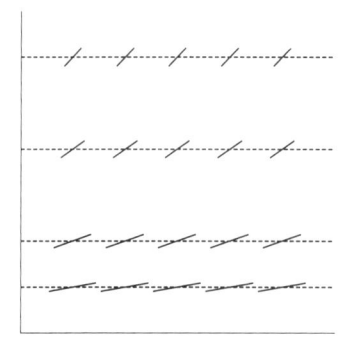

Figure 13.3. A direction field.

Exercise 13.4

Direction fields

Use the software to draw direction fields and superimpose a solution starting from some point of your choice (use the "Starting gun" tool). Notice that the solution can proceed forward or backward from that point.

Isoclines are simply curves that join points where $f(x,y)$ has the same value. They can be used to help visualize the direction field and hence the solution. The isocline for $f(x,y)=0$ is particularly significant, because the slope is 0 on it, and the solution curve must therefore have a stationary value there.

Note that for the simple growth model the isoclines are just straight lines parallel to the x-axis.

Example 13.2

Here are more examples of $f(x,y)$ to try out using the software.

$$x^2 + y^2 \qquad e^{-x^2} \qquad x^2 - y^2$$

$$\frac{x}{y} \qquad \frac{y}{x}$$

$$\sin(x^2 + y^2) \qquad \frac{y}{1-x}$$

Step-by-step Approach

Continuing using y for the dependent variable and x for the independent variable, we see that the simple population growth model has already been shown to take the form

$$\frac{dy}{dx} = ry.$$

How should we understand this? Return to the definition of a derivative. Suppose that we know the solution $y(x)$ and want to find the slope at a point x. We would consider:

$$\frac{y(x + \Delta x) - y(x)}{\Delta x}.$$

Instead of taking the limit, think of this as

$$\frac{\text{increment in } y}{\text{increment in } x}.$$

The differential equation in our example is telling us that in the limit in which $\Delta x \to 0$ this ratio is ry, so we have:

$$\text{increment in } y = \Delta y \cong ry\, \Delta x.$$

The "approximately equals" sign $\cong$ is needed because the ratio is only truly equal to ry in the limit.

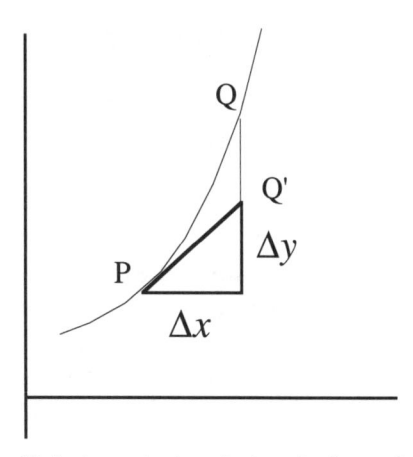

Figure 13.4. A step in the solution of a first order differential equation.

This statement can be interpreted graphically as shown in Figure 13.4. P and Q are two points on the exact solution curve. The point Q´ is the solution point predicted by assuming that the slope of the solution is constant in the interval Δx. The difference has, of course, been exaggerated in the diagram to emphasize the distinction, and as $\Delta x \to 0$ the error in making the approximation will become relatively smaller. This step-by-step way of thinking of a differential equation can be applied to any first order differential equation.

Exercise 13.5

Step-by-step solution

Using the software, choose starting values for y and x, then choose a small value for Δx, and use this to estimate a new value for y. Halve Δx and try again. Try out four or five values of Δx and convince yourself that the approximate solutions are getting closer to some limiting curve.

Separation of Variables

We have seen that the general first order differential equation has the form

$$\frac{dy}{dx} = f(x, y),$$

and we have seen several examples where a closed-form solution (also sometimes called an *analytic solution*) can be found. Unfortunately, most choices of $f(x,y)$ lead to equations that do not have an analytic solution, but it is very useful to have techniques for finding an analytic solution where possible.

We look at the case where $f(x,y)$ can be separated into two functions $g(x)$ and $h(y)$: that is, $f(x,y)=g(x)h(y)$.

Note that this includes the easiest case of all when $h(y)=a$ (constant), which is just the case in which $f(x,y)$ is a function of x only.

Then $\dfrac{dy}{dx} = ag(x)$,

$$\therefore y = a \int g(x) dx + C.$$

Even in this case it might not be possible to find an analytic solution. For example, there is no analytic form for $\int e^{-x^2} dx$.

Returning to the more general case with $f(x,y)=g(x).h(x)$ generally,

$$\frac{dy}{dx} = g(x)h(y)$$
$$\therefore \frac{1}{h(y)} \frac{dy}{dx} = g(x) dx.$$

Integrate both sides with respect to x and use the substitution rule for integrals to get

$$\int \frac{dy}{h(y)} = \int g(x) dx.$$

Further progress depends on whether the integrals can be found in closed form.

Example 13.3

Use separation of variables to solve the differential equation $\dfrac{dy}{dx} = xy$ with the initial condition $y(0)=1$.

In the general method, take $g(x)=x$ and $h(y)=y$. Then $\int \dfrac{dy}{y} = \int x dx$, and therefore $A + \ln(y) = B + \tfrac{1}{2} x^2$. Combine A and B into a single constant by letting C satisfy $\ln(C) = B - A$ to give $\ln(y) = \ln(C) + \tfrac{1}{2} x^2$ and therefore $y = Ce^{\frac{1}{2}x^2}$. Since $y(0)=1$, $C=1$ and we have the solution $y = e^{\frac{1}{2}x^2}$.

Example 13.4

Use separation of variables to solve the differential equation $\dfrac{dy}{dx} = (x+a)\sec(y)$ with the initial condition $y(0)= y_0$.

In the general method take $g(x)=x+a$ and $h(y)=\sec(y)$. Then $\therefore \int \dfrac{dy}{\sec(y)} = \int (x+a) dx$, and therefore $A + \sin(y) = B + \tfrac{1}{2}(x+a)^2$. Combine A and B into a single constant by letting $C = B - A$ to give $y = \arcsin(C + \tfrac{1}{2}(x+a)^2)$. Since $y(0)= y_0$, $C = \sin(y_0) - \tfrac{1}{2} a^2$.

Second Order ODEs

The first order ODE $\dfrac{dy}{dx} = f(x, y)$ can be thought of as an equation defining the first derivative of the solution at any point where x and the solution y are known. To solve it we need an initial condition.

$$ap^2 + bp + c = 0 \quad (13.11)$$

This quadratic has solutions given by

$$p = -\frac{b}{2a} \pm \frac{\sqrt{b^2 - 4ac}}{2a} \quad (13.12)$$

There are three cases to consider depending on whether $b^2 - 4ac$ is > 0, $= 0$, or < 0.

Case (i) $b^2 - 4ac > 0$

There are two real roots for p, and either of the functions

$$y = e^{p_1 x}, \; e^{p_2 x}$$

is a solution of (13.10). We have found not just one solution, but two. Because the equation is linear and homogeneous, we can see that if A and B are arbitrary constants then

$$y = Ae^{p_1 x} + Be^{p_2 x} \quad (13.13)$$

is also a solution. The two arbitrary constants allow us to fit the values of y and y' to the given initial conditions, so this is the general solution.

Case (ii) $b^2 - 4ac = 0$

There are two equal roots, $p = -\frac{b}{2a}$. However this alone would not allow us to fit the two initial conditions, so we look for another. Try $y = xe^{px}$. Differentiating twice with respect to x gives $y' = (1 + px)e^{px}$ and $y'' = (2p + p^2)e^{px}$ which satisfies the equation. Combining arbitrary multiples of these two solutions gives the general solution

$$y = (A + Bx)e^{px} \quad (13.14)$$

where $p = -\frac{b}{2a}$.

Case (iii) $b^2 - 4ac < 0$

The roots have square roots of negative numbers in them.

If you are familiar with complex numbers and Euler's Formula in particular, then you will recall that

$$e^{i\theta} = \cos\theta + i\sin\theta$$

and so we can write the general solution as

$$y(x) = e^{\alpha x}(A\cos(\omega x) + B\sin(\omega x)) \quad (13.15)$$

where $\alpha = -\dfrac{b}{2a}$, and $\omega = \dfrac{\sqrt{4ac - b^2}}{2a}$

If you are not familiar with complex numbers, then the general solution (13.15) can be established by verifying that the functions $Ae^{\alpha x}\cos(\omega x)$, $Be^{\alpha x}\sin(\omega x)$ satisfy the differential equation (13.10), with α and ω defined as in equation (13.15).

Exercise 13.7

Second order ODEs
Use the software to explore the characteristics of each of the above solutions. You can choose the values of a, b, and c as used in equation (13.10), and see the corresponding values of p_1 and p_2. You can also choose the initial values for x, y, and y', and see the corresponding values of A, B, and a graph of the solution.

Example 13.7
Solve the second order ODE $y'' + y' + y = 0$ with the initial conditions $y(0)=0$, $y'(0) = 1$. Make the substitution $y = e^{px}$. Then $p^2 + p + 1 = 0$ so $p = -\frac{1}{2} \pm i\frac{\sqrt{3}}{2}$. The general solution is therefore

$$y(x) = e^{-\frac{1}{2}x}\left(A\cos(\tfrac{\sqrt{3}}{2}x) + B\sin(\tfrac{\sqrt{3}}{2}x)\right).$$

The condition $y(0)=0$ gives $A=0$, and the condition $y'(0)=1$ gives $1 = B\frac{\sqrt{3}}{2}$. The required solution is therefore

$$y(x) = \tfrac{2}{\sqrt{3}} e^{-\tfrac{1}{2}x} \sin(\tfrac{\sqrt{3}}{2} x).$$

If complex numbers are not to be used, then try $y(x) = e^{\alpha x} \sin(\omega x)$. Differentiation gives

$$y'(x) = e^{\alpha x}(\alpha \sin(\omega x) + \omega \cos(\omega x))$$

and then

$$y''(x) = e^{\alpha x}\left((\alpha^2 - \omega^2)\sin(\omega x) + 2\alpha\omega \cos(\omega x)\right).$$

As $y'' + y' + y = 0$ we have

$$(\alpha^2 - \omega^2 + \alpha + 1)\sin(\omega x) + \omega(2\alpha + 1)\cos(\omega x) = 0.$$

For this expression to vanish for all values of x requires that the coefficients of $\sin(\omega x)$ and $\cos(\omega x)$ each vanish, and since $\omega \neq 0$ (or the solution would be always zero) we must have $\alpha = -\tfrac{1}{2}$ and therefore $\omega^2 = \tfrac{3}{4}$. This gives one solution as $y(x) = Be^{-\tfrac{1}{2}x} \sin(\tfrac{\sqrt{3}}{2}x)$. A similar calculation starting from $y(x) = e^{\alpha x} \cos(\omega x)$ shows that $y(x) = Ae^{-\tfrac{1}{2}x} \cos(\tfrac{\sqrt{3}}{2}x)$ is another possible solution. Hence we obtain the same solution as above.

Example 13.8
Solve the second order ODE $y'' + y = 0$ with the initial conditions $y(0)=0$, $y'(0)=1$. (The equation $y'' + y = 0$ is called the Simple Harmonic Equation.)

Make the substitution $y = e^{px}$. Then $p^2 + 1 = 0$ so $p = \pm i$. The general solution is therefore $y(x) = A\cos(x) + B\sin(x)$. The condition $y(0)=0$ gives $A=0$, and the condition $y'(0)=1$ gives $B=1$. The required solution is therefore $y(x) = \sin(x)$.

If complex numbers are not to be used, the technique used in the second part of Example 13.7 above can be used again and will give the solution $y(x) = \sin(x)$ as just found.

Phase Plane

The obvious way to plot graphs of solutions $y(x)$ is to plot the dependent variable $y(x)$ against the independent variable x. The software gives several examples of this type of plot. For example, the solution found in Example 13.8 is shown in Figure 13.6, which also shows the plot of $y'(x)$ against x superimposed.

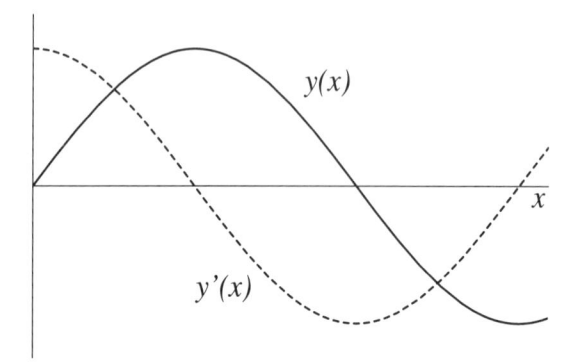

Figure 13.6. The solution of $y'' + y = 0$ with the initial conditions $y(0) = 0$, $y'(0) = 1$.

In a second order ODE, it is often useful to plot y' against y. This is called a *phase plane* plot. The solution found in Example 13.8 gives a circle. To see that this must be so, let $z = y'$. From the solution found, $y = \sin(x)$ and $z = \cos(x)$. Therefore, $y^2 + z^2 = 1$, which is the equation of a circle of unit radius centered on the origin. Note that x does not appear explicitly in the plot, but the solution at any given value of x will correspond to one point in the phase plane. In this case, as x increases, the point corresponding to a given x just moves around the circle periodically at a uniform rate with respect to x. This is demonstrated in the software, as suggested in the next exercise.

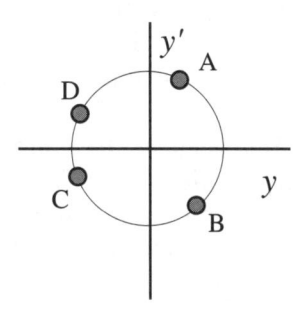

Figure 13.7. Phase plane plot for $y'' + y = 0$.

Exercise 13.8

Phase plane

Use the software to generate a phase plane plot of $y'' + y = 0$ for which, as we have seen, a solution is $y(x) = \sin(x)$, giving $y'(x) = \cos(x)$. Observe the following features. It may help to think of x as time, since in the (x,y) plane the solution is found by the computer, with x steadily increasing.

1. The point (y, y') moves in a clockwise direction. To see why this is so, ask yourself the following questions: When y' is positive, does y increase or decrease with respect to x? When (y, y') is at a point like A in Figure 13.7, is y increasing or decreasing with respect to x? What can you say about the direction of movement of (y, y') at points like B, C, and D?

2. The maxima and minima on the (x,y) plot as, for example, in Figure 13.6 correspond to points (y, y') which lie on the y-axis in the phase plane. To see why this is so, ask yourself the following questions: What can you say about y' when $y(x)$ is at a maximum or minimum? What can you say about y' for points in the (y, y') which lie on the y-axis?

3. Given a particular phase plane plot, the shape of the phase plane plot is not affected if a new starting point is chosen that lies on the given plot.

Phase plane plots will not always be circles, nor need they be closed curves, as the following exercise and example show.

Exercise 13.9

Phase plane plots

Use the software to generate phase plane plots of the following ODEs (and others of your choice) using parameter values and starting conditions of your choice.

$$y'' = a$$
$$y'' = ax$$
$$y'' + y' = 0$$
$$y'' + y = 0$$
$$y'' + ky' + p^2 y = 0$$
$$y'' + ky' + p^2 y = \sin(x)$$
$$y'' + ky'(y^2 - 1) + y = 0$$

For each equation, consider the following points when watching the solution develop.

1. In what direction does the point (y, y') move?

2. What points in the (y, y') plane correspond to maxima and minima in the (x,y) plane?

3. What happens if given a phase plane plot you choose new starting conditions corresponding to some point on the given plot?

Observe that for the Simple Harmonic Motion (SHM) equation $y'' + y = 0$ the phase plane plot is indeed an ellipse as shown in Exercise 13.9. What happens when a friction term is added, as with $y'' + ky' + p^2 y = 0$?

Foxes and Rabbits

Second and higher order ODEs can sometimes be more meaningfully written in the form of a system of equations.

Here is an example. Suppose there is an island with a population of x rabbits and y foxes (using units of $x=1$ to mean 10,000 rabbits, $y=1$ to mean 1000 foxes). If there are no foxes, the rabbit population increases according to the simple growth law $\frac{dx}{dt} = 2x$. On the other hand, the foxes will die off if there are no rabbits to eat according to the law $\frac{dy}{dt} = -y$. To model the interaction of the two species, suppose it is found that with foxes present, the rabbits' growth rate of 2 is modified to $2(1-y)$. With rabbits to eat, the foxes' growth rate becomes $(-1+x)$. We then get a model known as a Volterra System:

$$\left. \begin{array}{l} \frac{dx}{dt} = 2x(1-y) \\ \frac{dy}{dt} = (-1+x)y \end{array} \right\} \quad (13.16)$$

If the solution is plotted as $y(t)$ versus $x(t)$, the cyclic nature of the population is clearly seen. Note that the $y(t)$ versus $x(t)$ is not the same type of plot as the phase plane plot in the (y, y') plane as used in the previous section.

> **Exercise 13.10**
>
> **Foxes and Rabbits**
> Run the foxes and rabbits simulation. Try different starting points for the population.

Higher Order ODEs

The earlier sections of this chapter explain the basic features of first and second order ordinary differential equations and introduce a very few of the mathematical techniques for solving them. Standard textbooks contain much more detail on methods of solution and on properties of other important classes of ODE. However, we do not want to conclude this chapter without giving the reader a glimpse of higher order systems of ODEs, because these are so important in many applications. We will do this without presenting any detailed analytical treatment, relying on the power of computer solutions and graphics to bring out some qualitiative features.

First, it is useful to note that any ODE of order n can be written down as a system of n first order DEs. There has already been an example of a system of two first order ODEs (equation (13.16) in the foxes and rabbits example above), but systems of second and higher order can also be written in a similar way. For example, to express as a system of two first order equations, let $z(x) = y'(x)$. Then:

$$\left. \begin{array}{l} y' = z \\ z' = -y \end{array} \right\} \quad (13.17)$$

There will also be initial conditions for y and z.

In general, if the dependent variables are $y_1(x), y_2(x), \ldots, y_i(x), \ldots, y_n(x)$, then we will have

$$\left. \begin{array}{l} y'_1 = f_1(x, y_1, y_2, \ldots, y_n) \\ y'_2 = f_2(x, y_1, y_2, \ldots, y_n) \\ \ldots \\ y'_i = f_i(x, y_1, y_2, \ldots, y_n) \\ \ldots \\ y'_n = f_n(x, y_1, y_2, \ldots, y_n) \end{array} \right\} \quad (13.18)$$

plus initial conditions for $y_1, y_2, \ldots, y_n$.

For the example $y'' + y = 0$, take $y_1 = y$ and $y_2 = y'$, and then $f_1(x, y_1, y_2) = y_2$, $f_2(x, y_1, y_2) = -y_1$. This is exactly the same as equation (13.17) except that the variables have been renamed.

There are considerable problems in solving such systems in general, by whatever means, analytic or numerical. Modern algorithms can solve a wide variety with pre-determined accuracy.

Solutions can be very complicated, as demonstrated by the Lorentz System:

$$\left.\begin{array}{l} x' = -\sigma(x-y) \\ y' = rx - y - xz \\ z' = -\beta z + xy \end{array}\right\} \quad (13.19)$$

This is a third order system in variables $x(t), y(t), z(t)$ that originally arose in theoretical studies of the weather. One way of visualizing the solution is to plot x, y, z in three dimensions, as has been done in the solution of the system shown in Figure 13.8, which gives a solution with σ=3, r=26.5, β=1. This system exhibits what is called *chaotic behavior*.

> **Exercise 13.11**
>
> **The Lorentz System**
> Use the software to solve the Lorentz Equation.

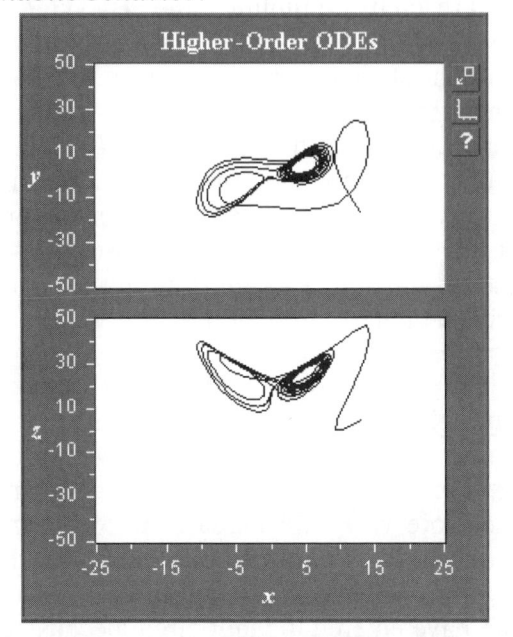

Figure 13.8. A solution of the Lorentz Equation.

WORKED EXAMPLE 13.1
FIRST ORDER ODE

Sketch direction fields for the differential equation
$$y' = x^2 + y^2 - 1.$$
Draw the isocline for zero slope, and identify the region where the slope is negative. Sketch in guesses for solution curves through the points
$$(-2, 0), \ (0, \tfrac{1}{2}), \ (0, 0).$$

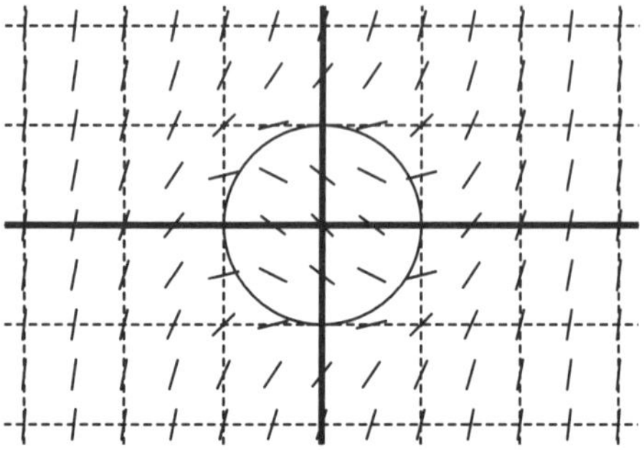

Figure 13.9. Direction field and isocline for zero slope.

The slope is a function whose magnitude increases as the magnitudes of x and y increase (See Figure 13.9). Along the line $y=0$, for example, the slope is given by the function $x^2 - 1$, which is zero at $x = \pm 1$ and increases without limit as $x \to \pm\infty$. Because x^2 is always positive, it is clear that $x^2 - 1$ has a minimum value of -1 at $x=0$ and is negative in the interval $[-1, 1]$. We can make similar statements about the way the slope varies along the y-axis.

Since zero is of special interest as a value for the slope, we next ask at what other points the slope is zero, and from the differential equation itself we see that these points must satisy
$$0 = x^2 + y^2 - 1.$$

When we rewrite this in the form
$$x^2 + y^2 = 1$$
we get the equation of the unit circle with the center at the origin. This circle is therefore the isocline for zero slope. Inside this circle, the slope is negative, and the origin is the point where it takes its minimum value of -1. Outside the circle, the slope is always positive. Note the symmetry in both x and y.

A solution curve that intersects the unit circle will have a maximum, minimum, or point of inflection at the point of intersection, because we know that the slope being zero is a condition for these stationary points. Such a solution curve must have a portion inside the unit circle unless the intersection is at $(0, 1)$ or $(0, -1)$. If there is a portion inside the unit circle, then working in the direction of increasing x, such a curve will enter the unit circle at an intersection where $x < 0$. Then it will acquire a negative slope, which will become more negative as x increases until $x=0$ where the slope will reach its minimum value. By symmetry it will then follow the reverse pattern, with the slope becoming less negative until an interesection at a positive x-value is reached where the curve leaves the unit circle and from then on takes increasing positive values of slope. The intersections at $(0,1)$ or $(0,-1)$ are a special case: the solution will touch the unit circle here and have a point of inflection.

Clearly, most solution curves will not go through the unit circle at all. Any curve going through $(0, y)$ for which $y > 1$ (for example) will have positive slope, and therefore $y(x)$ will increase for $x > 0$ and move further from the unit circle. For the region $x < 0$, we know that the curve cannot have entered the unit circle because of our analysis above of curves that do enter.

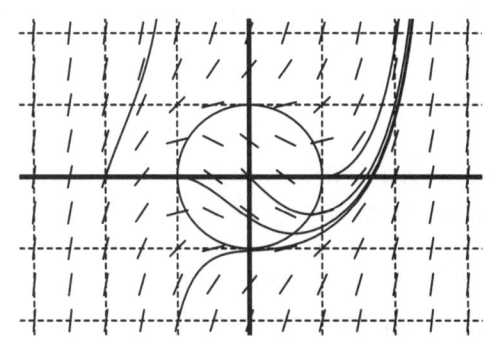

Figure 13.10. Direction field and solution of a first order ODE.

Combining these observations, we can sketch in some solution curves as shown in Figure 13.10.

WORKED EXAMPLE 13.2
BUNGEE JUMP

Find a solution to the elastic phase of motion of a bungee jump (or the first elastic phase if there are several).

It was shown earlier in this chapter that the equation of motion for a bungee jump may be written as

$$\frac{d^2x}{dt^2} + k\frac{dx}{dt} + \omega^2 x = g + \omega^2 L \quad (13.7)$$

where $\omega^2 = \frac{\lambda}{mL}$, with initial conditions $x=L$ at $t=0$, and $x' = \frac{dx}{dt} = v_0$ at $t=0$. Recall that here x is measured downward from the point from which the bungee is suspended, and that friction is assumed to be given by a force mkv in the direction opposite to the motion.

The elastic phase begins as soon as the tension in the bungee becomes greater than zero, which happens as soon as $x > L$. A general solution is known for equations of form (13.10), but (13.7) is not quite of this form because there is a nonzero term on the right-hand side. Let

$$z = x - \left(\frac{g}{\omega^2} + L\right) \quad (13.20)$$

Substitute

$x = z + X$, where $X = \left(\frac{g}{\omega^2} + L\right)$ (13.21)

into (13.7) to get

$$\frac{d^2z}{dt^2} + k\frac{dz}{dt} + \omega^2 z = 0 \quad (13.22)$$

Notice that the constant terms have canceled out. This happens because X is a constant: therefore, we get $\frac{dx}{dt} = \frac{dz}{dt}$ and $\frac{d^2x}{dt^2} = \frac{d^2z}{dt^2}$, and when $x = z + X$ is substituted into the third term on the left of (13.7), $\omega^2 x = \omega^2 z + \omega^2 X$ and $\omega^2 X$ cancels with the constant on the right-hand side of (13.7).

Note that $z=0$ is a valid solution of (13.22), representing an equilibrium in which the bungee jumper has come to rest. This gives the equilibrium position of the jumper as

$x = \left(\frac{g}{\omega^2} + L\right)$ below the suspension point of the bungee, that is, the natural length L of the bungee plus enough stretching to give sufficient tension to support the jumper's weight.

Now apply the general solution to (13.10) as found in the earlier section, remembering that the independent variable is now t rather than x. In terms of the constants a, b, and c used, identify:

$$a=1, \quad b=k, \quad c=\omega^2.$$

Case (i) $k^2 - 4\omega^2 > 0$

$$z = e^{p_1 t}, \; e^{p_2 t}$$

where $p_1, p_2 = \frac{1}{2}\left(-k \pm \sqrt{k^2 - 4\omega^2}\right)$ (13.23)

Therefore,
$$x = X + Ae^{p_1 t} + Be^{p_2 t} \quad (13.24)$$

Note that p_1 and p_2 are both negative, so that $x \to X$ as $t \to 0$.

Case (ii) $k^2 - 4\omega^2 = 0$

The two roots are equal, $p_1, p_2 = -\frac{1}{2}k$, and we get
$$x = X + (A + Bt)e^{-\frac{1}{2}kt} \quad (13.25)$$

Case (iii) $k^2 - 4\omega^2 < 0$

The two roots are complex and the form of the solution changes to
$$x = X + e^{-\frac{1}{2}kt}(A\cos(\nu t) + B\sin(\nu t)) \quad (13.26)$$
where $\nu = \sqrt{\omega^2 - \frac{1}{4}k^2}$.

The initial conditions can be applied to these forms. Alternatively, work in terms of z using $z = L - X$ at $t=0$, and $z' = \dfrac{dz}{dt} = v_0$ at $t=0$.

WORKED EXAMPLE 13.3
GENERAL PHASE PLOT FOR SHM

Solve the second order ODE
$$\frac{d^2y}{dt^2} + \omega^2 y = 0$$
with the initial conditions
$$y(0) = y_0, \quad y'(0) = y'_0$$
and determine the equation of its phase plane plot (i.e., the curve traced by the point (y, y')). This is similar to Example 13.7, but uses t as the independent variable.

Make the substitution $y = e^{pt}$. Substituting in the differential equation gives

$p^2 + \omega^2 = 0$, so $p = \pm i\omega$. The general solution is therefore

$$y(t) = A\cos(\omega t) + B\sin(\omega t).$$

This gives $y(0) = A$ and $y'(0) = B\omega$. The condition $y(0) = y_0$ gives $A = y_0$, and the condition $y'(0) = y'_0$ gives $B = \dfrac{y'_0}{\omega}$. The required solution is therefore

$$y(t) = y_0 \cos(\omega t) + \left(\frac{y'_0}{\omega}\right)\sin(\omega t).$$

To determine the shape of the phase plane plot, we have $y(t) = A\cos(\omega t) + B\sin(\omega t)$ and $y'(t) = -A\omega\sin(\omega t) + B\omega\cos(\omega t)$.

Therefore,
$$y^2 + \frac{y'^2}{\omega^2} =$$
$$(A^2 + B^2)\cos^2(\omega t) + (A^2 + B^2)\sin^2(\omega t)$$
$$= (A^2 + B^2).$$

Recall that in the (y,z) plane an ellipse with semi-axes a, b has equation $\dfrac{y^2}{a^2} + \dfrac{z^2}{b^2} = 1$, so we can identify the phase plane curve as an ellipse with semi-axis length $\sqrt{y_0^2 + \dfrac{y'^2_0}{\omega^2}}$ on the y-axis and $\sqrt{\omega^2 y_0^2 + y'^2_0}$ along the y'-axis. Note that when $\omega = 1$ these are equal and we get a circle as before.

PROBLEM 13.1
FIRST ORDER EQUATIONS

Name: _____
Date: _____
Section: _____

Problem 13.1a
Sketch direction fields for each of the following differential equations, and then sketch in a guess for solution curves through the given points.

(i) $\quad y' = e^{-x^2}$

What is the isocline for slope 1? Is the slope ever negative?

Sketch guesses for solution through: (0, 0), (0, 1), (0, -1).

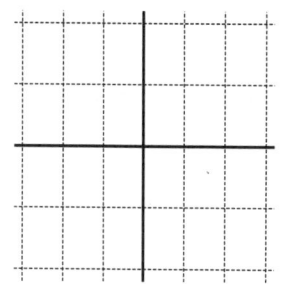

(ii) $\quad y' = x^2 - y^2$

Draw the isoclines for zero slope, and identify the regions where the slope is negative.

Sketch a guess for solution through: (0, -1), (0, 1), (-1, 0), (1, 0).

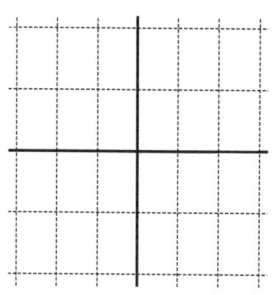

(iii) $\quad y' = \dfrac{x}{y}$

What is the isocline for zero slope? Identify the regions where the slope is negative. What can you say about the line $y=0$?

Sketch a guess for solution through: (0, 1), (0, -1), (1, 1).

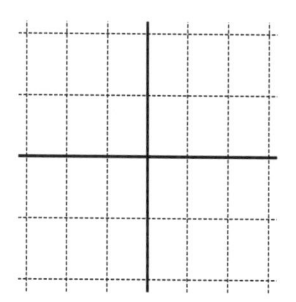

(iv) $\quad y' = \dfrac{y}{x}$

What is the isocline for zero slope? Identify the regions where the slope is negative. What can you say about the line $x=0$?

Sketch a guess for solution through: (-1, 0), (1, 0), (1, 1).

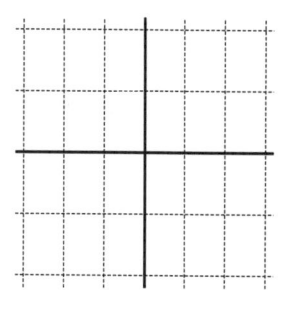

PROBLEM 13.1
FIRST ORDER EQUATIONS

(Continued)

(v) $y' = xy$

What are the isoclines for zero slope? Identify the regions where the slope is negative.

Sketch a guess for solution through: (1, 1), (1, -1), (-1, -1), (-1, 1).

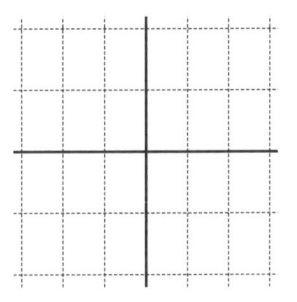

Problems 13.1b

Use the method of separation of variables to obtain general closed-form solutions of the following differential equations. Obtain values for the constants which give a solution curve through the points given. Check that these agree with your sketches made in Problems 13.1a (use the graph software if you find that helpful).

(vi) $y' = \dfrac{x}{y}$ Find the solution through: (0, 1), (0, -1), (1, 1).

(vii) $y' = \dfrac{y}{x}$ Find the solution through: (-1, 0), (1, 0), (1, 1).

(viii) $y' = \dfrac{y}{x}$ Find the solution through: (-1, 0), (1, 0), (1, 1).

(ix) $y' = xy$ Find the solution through: (1, 1), (1, -1), (-1, -1), (-1, 1).

PROBLEM 13.2
SECOND ORDER ODES

Name: _____
Date: _____
Section: _____

Problem 13.2

(i) A cautious novice bungee jumper decides to jump from a height at which the bungee is just about to stop being slack. In terms of the notation used in Worked Example 13.2, this is equivalent to taking initial conditions $x=L$ at $t=0$, and $x' = \dfrac{dx}{dt} = 0$ at $t=0$. Neglecting air resistance (take $k=0$), show that case (iii) (as in Worked Example 13.2) is the appropriate form of the solution. Show that the lowest height reached (i.e., the maximum x) corresponds to $x = L + 2X$, and find the maximum speed attained during the jump. Describe the motion predicted by the mathematics in this case (taking $k=0$). Does the jumper ever come to rest?

(ii) Try to repeat (i) but taking account of air resistance (i.e. with $k \neq 0$). What is the condition for the appropriate form of solution to be $x = X + e^{-\frac{1}{2}kt}(A\cos(vt) + B\sin(vt))$ where $v = \sqrt{\omega^2 - \frac{1}{4}k^2}$? Assuming that this condition holds, and that the same initial conditions apply as in (i), show that $A = -\dfrac{g}{\omega^2}$ and $B = \dfrac{kA}{2v}$. Find the time at which $x'=0$ for the first time after the jump starts. Hence find an expression for the minimum height reached during the jump. Verify that in the limit $k \to 0$ this gives the same solution as found in (i). Describe the motion predicted by the mathematics in this case (taking $k \neq 0$). Does the jumper ever come to rest? When is the lowest height of the entire jump reached?

PROBLEM 13.2
SECOND ORDER ODES
(Continued)

(iii) A middle-aged bungee jumper likes the free-fall phase but dislikes bouncing up and down. The jumper also wants the elastic phase to be over in the shortest possible time. Supposing that k is determined by the shape of the jumper (and that nothing short-term can be done about this), and that the jumper's preferences are to be met by ordering a special bungee with a specified modulus of elasticity, what modulus of elasticity would you advise (in terms of variables as defined in Worked Example 13.2)? (Hint: Show that case (ii) is the appropriate solution.) Will the jumper ever come to rest under this model? How do you justify your advice as meeting the requirement that the elastic phase be over in the shortest possible time?

(iv) Using methods similar to those of Worked Example 13.3, investigate the phase plane curve of $\dfrac{d^2x}{dt^2} + k\dfrac{dx}{dt} + \omega^2 x = 0$ in the three cases $k^2 - 4\omega^2 > 0$, $=0$, and < 0. Give sketches and justify your results. You may assume any suitable starting conditions (except $x = x' = 0$, of course).

NEW SITUATIONS

Forced oscillations

As indicated in this chapter, only a very few methods for solving ODEs have been mentioned. Here is a simple but important extension of the method described in the section Linear Second Order ODEs. Using t rather than x as the independent variable, instead of solving

$$a\frac{d^2y}{dt^2} + b\frac{dy}{dt} + c = 0 \qquad (13.27)$$

we shall consider

$$a\frac{d^2y}{dt^2} + b\frac{dy}{dt} + c = F\cos(ft+\phi) \qquad (13.28)$$

This equation can be used to describe a system in which there are forced oscillations. Equation (13.27) on its own allows solutions that are zero for all t, whereas (13.28) clearly does not.

Note first that if $y_H(t)$ satisfies the homogeneous equation (13.27) and $Y(t)$ satisfies (13.28), then $Y(t) + Cy_H(y)$ also satisfies (13.28) for any arbitrary constant C. Since we already know the general solution of (13.27), a general solution to (13.28) can be found if we can find just one function satisfying (13.28). Such a solution is known as a *particular integral*.

Finding particular integrals is often a matter of making an informed guess. In this case we know that the first and second differentials of a combination of cos and sin functions, for example,

$$Y(t) = \alpha\cos(ft+\phi) + \beta\sin(ft+\phi) \qquad (13.29)$$

will also be combinations of cos and sin functions, so there is a good prospect that suitable choices of α and β will allow us to satisfy the equation. Without giving all the intermediate details (which you can work out on paper or with the help of a computer algebra package), differentiating (13.29) twice, then substituting in (13.28) gives

$$\{-a\alpha f^2 + b\beta f + c\alpha\}\cos(ft+\phi) +$$
$$\{-a\beta f^2 - b\alpha f + c\beta\}\sin(ft+\phi) = F\cos(ft+\phi).$$

To make the solution work therefore requires:

$$\{-a\alpha f^2 + b\beta f + c\alpha\} = F$$
$$\{-a\beta f^2 - b\alpha f + c\beta\} = 0$$

Solving for α and β then gives

$$\alpha = \frac{(c - af^2)F}{D}$$

$$\beta = \frac{bfF}{D} \qquad (13.30)$$

with $D = \left[b^2f^2 + (af^2 - c)^2\right]$.

It is also useful to note that
$$Y(t) = \alpha\cos(ft+\phi) + \beta\sin(ft+\phi)$$
may be written as
$$Y(t) = R\cos(ft+\phi+\varphi) \qquad (13.31)$$
where $R = \sqrt{\alpha^2 + \beta^2}$ and $\tan(\varphi) = \frac{-\beta}{\alpha}$.

From (13.30) it can be shown that $R = F/D$ in this case. The interpretation is that the solution forced by the term on the right-hand side is also sinusoidal, but with amplitude multiplied by the factor F/D and phase shifted by φ.

If you are familiar with complex numbers, then it is worth noting that (13.31) can be obtained easily by regarding the right-hand side of (13.28) as $\mathrm{Re}[e^{i(ft+\phi)}]$ and looking for a solution of form

$$Y(t) = \mathrm{Re}[\rho e^{i(ft+\phi)}].$$

It will be found that $\rho = R\,e^{i\varphi}$.

Chapter 14

Spherical and Polar Coordinates

Prerequisites

Before you study this material, you should be familiar with:

(1) Functions and their graphs (Module 1).

(2) The elements of differential and integral calculus (Modules 2, 3, 5, 6, and 7).

Objectives

This chapter introduces polar coordinates (r, θ) as an alternative to Cartesian coordinates (x, y) for describing points in two dimensional space, and cylindrical polars (r, θ, z) and spherical polars (ρ, θ, ϕ) as alternatives to (x, y, z) for points in three dimensions.

If there are to be mathematical theories to explain and model physical and scientific phenomena generally, then we need mathematical ways of describing positions in space. This is what coordinate systems are. However there are many different ways of describing positions in space. The familiar Cartesian system, which is in many ways like the map grid reference system, is not always the most convenient. We will see that because any point can be described equivalently in any coordinate system, there must be ways of converting from one system to another. This chapter explains how to do that.

Connections

Radar (*RA*dio *D*etecting *A*nd *R*anging) scanners do not measure (x, y) grid references; for surface use, as for shipping in the first video clip and simulation in the software, scanners measure range and bearing. (See Figure 14.1.) Range is found by timing pulses of radio waves transmitted by the radar antenna. The bearing is taken from the current angle of rotation of the scanner.

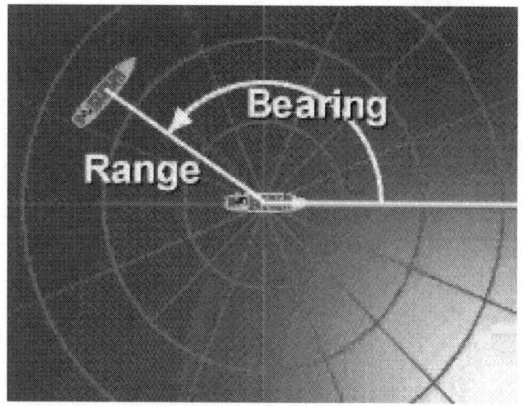

Figure 14.1. A radar screen.

Notice that all points with the same bearing lie on a radial line through the origin, and that all points with the same range lie on circles around the origin.

When radar is used to track aircraft or objects in space, two measurements are no longer enough. For aircraft, the height also needs to be known. The second video shows this situation. In mathematics, there are two commonly used coordinate systems in addition to the Cartesian that bring in the third dimension: they are cylindrical and spherical polar coordinates.

Plane Polar Coordinates

Polar coordinates used in mathematics are just like range and bearing used in radar. In mathematical notation, the symbols often used are r for range and θ (theta) for bearing. They are usually written in brackets as (r, θ). See Figure 14.2.

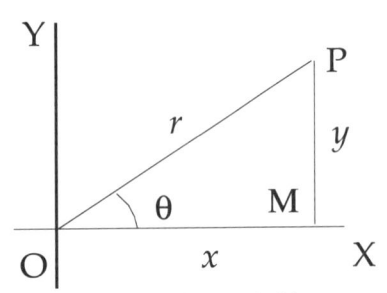

Figure 14.2. Polar coordinates (r, θ).

In a plane, the position of a point P relative to an origin O and a coordinate axis OX may be specified by the distance r of P from O (i.e., the distance OP), together with the angle POX (usually called θ). We will use radians for all angles, unless indicated otherwise.

Notice that for mathematical uses a bearing of zero is like east on a map, that is, along the x-axis when using Cartesian coordinates, and that the angle θ is measured positive in the anticlockwise direction.

By completing a right-angled triangle OPM where M is on the x-axis and OMP is a right angle, it can be seen that the Cartesian coordinates (x, y) of P are given by

$$x = r\cos(\theta), \quad y = r\sin(\theta) \quad (14.1)$$

This operation is called *transforming from polars to Cartesians*, or *converting from polars to Cartesians*. We can transform (or convert) from Cartesian coordinates to polar coordinates by using the inverse of this relationship, which is

$$r = \sqrt{x^2 + y^2}, \quad \theta = \arctan\left(\frac{y}{x}\right) + n\pi \quad (14.2)$$

where n is an integer chosen so that θ corresponds to the same direction as (x, y).

The reason that we cannot write just $\theta = \arctan\left(\frac{y}{x}\right)$ is that θ would then lie only in the interval $(-\frac{\pi}{2}, \frac{\pi}{2})$, the interval giving the principal values of the arctan function. Since $\tan(\theta) = \tan(\theta \pm n\pi)$, all possible values of θ satisfying $\tan(\theta) = \frac{y}{x}$ are given by $\arctan\left(\frac{y}{x}\right) + n\pi$ where n is any integer. The definition given in (14.2) can be applied to give θ in any standard interval.

Example 14.1
Transform the following points from polar coordinates to Cartesians:

$(r, \theta) = (1, \frac{\pi}{4}), (2, \frac{3\pi}{4}), (1, -\frac{\pi}{3}), (4, -\frac{2\pi}{3})$.

$\cos(\frac{\pi}{4}) = \sin(\frac{\pi}{4}) = \frac{1}{\sqrt{2}}$.

Therefore, $(r, \theta) = (1, \frac{\pi}{4})$ transforms to
$(x, y) = (\frac{1}{\sqrt{2}}, \frac{1}{\sqrt{2}})$.

$\cos(\frac{3\pi}{4}) = -\frac{1}{\sqrt{2}}, \sin(\frac{3\pi}{4}) = \frac{1}{\sqrt{2}}$.

Therefore, $(r, \theta) = (2, \frac{3\pi}{4})$ transforms to
$(x, y) = (-\sqrt{2}, \sqrt{2})$.

$\cos(-\frac{\pi}{3}) = \frac{1}{2}, \sin(-\frac{\pi}{3}) = -\frac{\sqrt{3}}{2}$.

Therefore, $(r, \theta) = (1, -\frac{\pi}{3})$ transforms to
$(x, y) = (\frac{1}{2}, -\frac{\sqrt{3}}{2})$.

$\cos(-\frac{2\pi}{3}) = -\frac{1}{2}, \sin(-\frac{2\pi}{3}) = -\frac{\sqrt{3}}{2}$.

Therefore, $(r, \theta) = (4, -\frac{2\pi}{3})$ transforms to
$(x, y) = (-2, -2\sqrt{3})$.

Example 14.2

Transform the following points from Cartesian coordinates to polars, expressing polar angles in the interval $-\pi < \theta \leq \pi$:

$(x, y) = (0, 2), (-1, 0), (\frac{\sqrt{3}}{2}, \frac{1}{2}), (-\sqrt{2}, -\sqrt{2})$.

The point $(x, y) = (0, 2)$ lies on the positive y-axis and therefore, $\theta = \frac{\pi}{2}$. The distance from the origin is 2. So $(r, \theta) = (2, \frac{\pi}{2})$.

The point $(x, y) = (-1, 0)$ lies on the negative x-axis and therefore, $\theta = \pi$. The distance from the origin is 1. Therefore, $(r, \theta) = (1, \pi)$.

For the point $(x, y) = (\frac{\sqrt{3}}{2}, \frac{1}{2})$, $\tan(\theta) = \frac{1}{\sqrt{3}}$. Now $\arctan\left(\frac{1}{\sqrt{3}}\right) = \frac{\pi}{6}$, and since the point lies in the quadrant $x > 0$, $y > 0$, we take $\theta = \frac{\pi}{6}$. We also have $r = \sqrt{x^2 + y^2} = 1$. Therefore, $(r, \theta) = (1, \frac{\pi}{6})$.

For $(x, y) = (-\sqrt{2}, -\sqrt{2})$, $\tan(\theta) = 1$. Now $\arctan(1) = \frac{\pi}{4}$, and since the point lies in the quadrant $x < 0$, $y < 0$, subtract π to give $\theta = -\frac{3\pi}{4}$. We also have $r = \sqrt{x^2 + y^2}$ which gives $r = 2$. Therefore, $(r, \theta) = (2, -\frac{3\pi}{4})$.

Exercise 14.1

Cartesians and polars

Use graph paper or a rough sketch to plot all the points given in Examples 14.1 and 14.2.

Exercise 14.2

Cartesians and polars

Use the software to explore the relation between Cartesians and polars. In particular, verify the conversions between the systems shown in Examples 14.1 and 14.2.

Plotting in Polars

Plotting (x, y) graphs of functions using relations like $y=y(x)$ is a frequent and familiar operation. The value of one coordinate (y in this case) is given as a function of the other (x in this case). The technique may be extended for use in any coordinate system. In polar coordinates, for example, r could be given as a function of θ, that is, $r=r(\theta)$. We expect that this will represent a curve in the plane. For example, $r=a$ (where a in a constant) simply gives a circle of radius a.

Example 14.3

Plot the function $r=|1+2\cos(\theta)|$ as a polar plot.

First note that $\cos(\theta)$ is periodic and that there is no need to consider values of θ outside the interval $-\pi < \theta \leq \pi$. Next note that $\cos(\theta)$ is symmetric in θ; therefore, the curve will be symmetric about the x-axis, and we need only work out what happens for the interval $0 < \theta \leq \pi$. Taking a few values for θ gives the following points that help us to sketch the curve.

(r, θ): $(3, 0)$, $(1+\sqrt{3}, \frac{\pi}{6})$, $(1+\sqrt{2}, \frac{\pi}{4})$, $(2, \frac{\pi}{3})$, $(1, \frac{\pi}{2})$, $(0, \frac{2\pi}{3})$, $(\sqrt{2}-1, \frac{3\pi}{4})$, $(1, \pi)$.

Since $\cos(\theta)$ is a decreasing function of θ in this interval, the point (r, θ) will be farthest from the origin at $\theta=0$ (where $r=3$) and get closer to the origin until θ reaches $\frac{2\pi}{3}$ where $\cos(\theta)=-\frac{1}{2}$ and therefore $r=0$. On the interval $\frac{2\pi}{3} < \theta < \pi$, r will increase until $r=1$ when $\theta=\pi$.

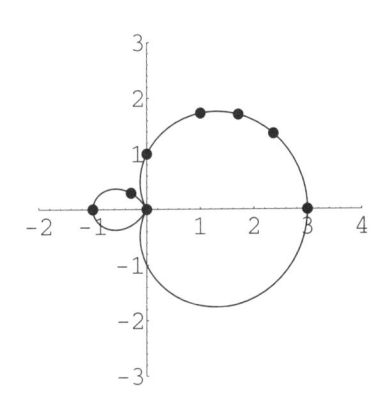

Figure 14.3. Polar plot of r=|1+2cos(θ)|.

Exercise 14.3

Plotting in polars
Use the software to plot r=|1+2cos(θ)|. Verify the sketch shown in Figure 14.3.

It will be noticed that polar plots in the form r=r(θ) have different characteristics from Cartesian plots like y=y(x). For example, the curve shown in Figure 14.3 has loops; there can be no choice of single-valued function[1] y(x) for which the graph given by y=y(x) is identical to the whole curve shown, simply because there would have to be at least two values of y for any values of x corresponding to points on the curve, and thus y(x) would not be a function.

A polar plot r=r(θ), where r(θ) is a single-valued function, can only give at most one value of r to each value of θ. However, this does not necessarily mean that a polar plot has at most one point in any direction, because θ itself may take infinitely many values for any one direction, unless the domain of the function r(θ) is restricted. For example, the values $\theta = \frac{\pi}{4}$, $\frac{9\pi}{4}$ both correspond to a direction of 45° to the x-axis, and so does any value of $\theta = (2n + \frac{1}{4})\pi$, n any integer. Figure 14.4 shows the curve r= $\frac{\theta}{2\pi}$ for $0 \leq \theta \leq 8\pi$.

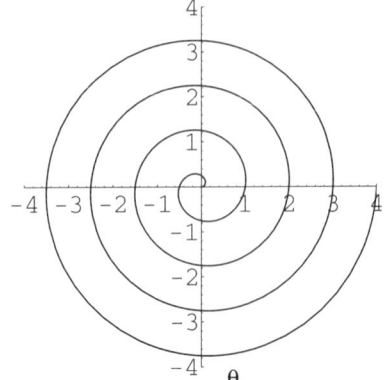

Figure 14.4. Polar plot of r= $\frac{\theta}{2\pi}$ *for* $0 \leq \theta \leq 8\pi$.

Transforming functions

Provided care is taken to deal with segments of curves where functions are single-valued, it is possible to convert the polar equation of a curve into a Cartesian equation, and viceversa. To convert from polars to Cartesian, substitute for r and θ using equation (14.2). To convert from Cartesian to polars, substitute for x and y using equation (14.1).

Example 14.4
Find the Cartesian form of the equation of a circle r=a.

Substitute for r using equation (14.2), giving:

$$r = \sqrt{x^2 + y^2} = a.$$

Squaring gives
$$x^2 + y^2 = a^2$$

---
[1] The qualifier *single-valued* has been used for emphasis. The definition of a function used here already requires that a function be single-valued.

which is the familiar Cartesian form of a circle with center at the origin. Notice that no problems arise with multiple values of y, because the equation has been obtained in implicit form. To write this curve in explicit Cartesian form, we could divide it into two semicircles:

$$y = \sqrt{a^2 - x^2}, \quad y = -\sqrt{a^2 - x^2}.$$

Example 14.5
Find the polar form of the equation of the parabola $y = x^2$.

Substitute for x and y using equation (14.1), giving $r\sin(\theta) = r^2 \cos^2(\theta)$. This can be rearranged as

$$r = \frac{\sin(\theta)}{\cos^2(\theta)}$$

where $0 \leq \theta < \frac{\pi}{2}$ or $\frac{\pi}{2} < \theta < \pi$.

Slopes

On the graph obtained by plotting $y=y(x)$, the slope of a tangent at a general point on the graph is given by the derivative $y'(x)$. This section shows how to obtain the slope of a tangent at a general point on the graph obtained by plotting $r=r(\theta)$.

Figure 14.5 shows a section of a curve $r=r(\theta)$ with two neighboring points P and Q, whose θ-values are $\Delta\theta$ apart. The slope will be found as

$$\lim_{\Delta\theta \to 0} \frac{\Delta y}{\Delta x}.$$

From equation (14.1), $x(\theta)=r\cos(\theta)$, so using the product rule for differentiation

$$\frac{dx}{d\theta} = r'(\theta)\cos(\theta) - r(\theta)\sin(\theta),$$

where $r' = \frac{dr}{d\theta}$. Similarly, $y(\theta)=r\sin(\theta)$, so

$$\frac{dy}{d\theta} = r'(\theta)\sin(\theta) + r(\theta)\cos(\theta).$$

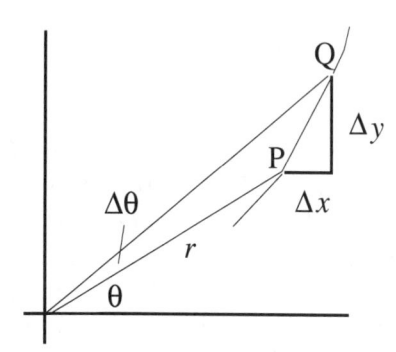

Figure 14.5. Slope of a polar plot.

For small $\Delta\theta$ we have

$$\Delta x \approx \frac{dx}{d\theta}\Delta\theta \text{ and } \Delta y \approx \frac{dy}{d\theta}\Delta\theta.$$

Therefore

$$\lim_{\Delta\theta \to 0} \frac{\Delta y}{\Delta x} = \frac{r'(\theta)\sin(\theta) + r(\theta)\cos(\theta)}{r'(\theta)\cos(\theta) - r(\theta)\sin(\theta)} \quad (14.3)$$

provided that $\frac{dx}{d\theta} \neq 0$ at (r, θ).

Example 14.6
Find the slope of a circle $r=a$ at any value of θ in the interval $[-\pi, \pi]$.

Since r is a constant, $r'(\theta) = 0$. Equation (14.3) gives

$$\text{slope} = -\cot(\theta).$$

This result may be verified by noting that the vector from the center of the circle to the point (r, θ) has slope $\tan(\theta)$, and the product of the slopes of two perpendicular directions is -1. See Figure 14.6.

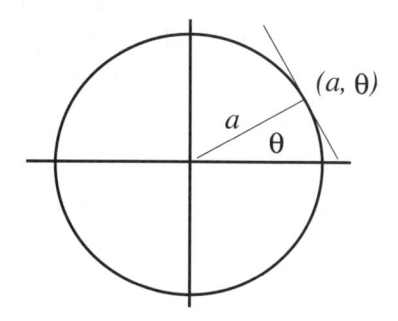

Figure 14.6. Slope of a circular arc.

Cylindrical Polar Coordinates

In three dimensional space (3-D) the position of a point P relative to an origin O and Cartesian axes OX, OY, OZ may be specified by *cylindrical coordinates* (r, θ, z). The z coordinate is the same as the Cartesian z coordinate and gives the displacement of P along the z-axis (See Figure 14.7). If P' is the projection of P onto the xy-plane, then r and θ are the plane polar coordinates of the point P' in the xy-plane. Thus, r is the length of OP', and θ is the angle XOP'.

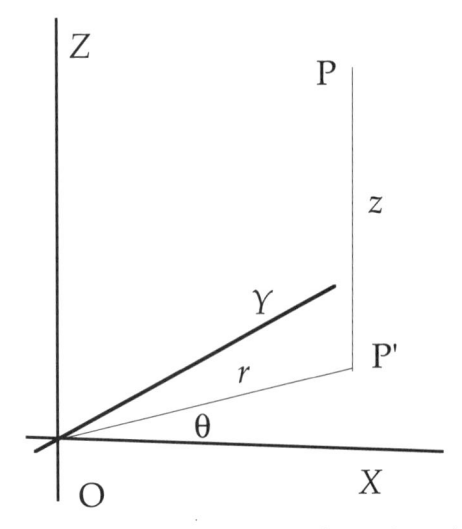

Figure 14.7. Cylindrical polar coordinates (r, q, z).

Cylindrical polar coordinates (r, θ, z) are very similar to plane polars except that the z coordinate is added. The Cartesian coordinates (x, y, z) of P are given by

$$x = r\cos(\theta), \quad y = r\sin(\theta), \quad z = z \quad (14.4)$$

We can transform Cartesian coordinates to cylindrical polar coordinates by noting that

$$\left.\begin{array}{l} r = \sqrt{x^2 + y^2} \\ \theta = \arctan\left(\dfrac{y}{x}\right) + n\pi \end{array}\right\} \quad (14.5)$$

where n is an integer that must be chosen in the same way as in equation (14.2) for plane polar coordinates. The z coordinate is the same in both systems.

Example 14.7
Transform the following points from cylindrical polar coordinates to Cartesians:

$(r, \theta, z) = (1, \frac{\pi}{4}, 0)$, $(2, \frac{3\pi}{4}, 2)$,

$(1, -\frac{\pi}{3}, -1)$, $(4, -\frac{2\pi}{3}, 1)$.

$\cos(\frac{\pi}{4}) = \sin(\frac{\pi}{4}) = \frac{1}{\sqrt{2}}$.

Therefore, $(r, \theta, z) = (1, \frac{\pi}{4}, 0)$ transforms to
$$(x, y, z) = (\tfrac{1}{\sqrt{2}}, \tfrac{1}{\sqrt{2}}, 0).$$

$\cos(\frac{3\pi}{4}) = -\frac{1}{\sqrt{2}}$, $\sin(\frac{3\pi}{4}) = \frac{1}{\sqrt{2}}$.

Therefore, $(r, \theta, z) = (2, \frac{3\pi}{4}, 2)$ transforms to
$$(x, y, z) = (-\sqrt{2}, \sqrt{2}, 2).$$

$\cos(-\frac{\pi}{3}) = \frac{1}{2}$, $\sin(-\frac{\pi}{3}) = -\frac{\sqrt{3}}{2}$.

Therefore, $(r, \theta, z) = (1, -\frac{\pi}{3}, -1)$ transforms to
$$(x, y, z) = (\tfrac{1}{2}, -\tfrac{\sqrt{3}}{2}, -1).$$

$\cos(-\frac{2\pi}{3}) = -\frac{1}{2}$, $\sin(-\frac{2\pi}{3}) = -\frac{\sqrt{3}}{2}$.

Therefore, $(r, \theta, z) = (4, -\frac{2\pi}{3}, 1)$ transforms to
$$(x, y, z) = (-2, -2\sqrt{3}, 1).$$

Example 14.8
Transform the following points from Cartesian coordinates to cylindrical polars, expressing polar angles in the interval $-\pi < \theta \leq \pi$:

$(x, y, z) = (2, 0, -1)$, $(-1, 0, 0)$,

$(\tfrac{1}{2}, -\tfrac{\sqrt{3}}{2}, -2)$, $(-\sqrt{2}, \sqrt{2}, 1)$.

The projection of (2, 0, -1) onto the xy-plane is (2, 0, 0) which lies on the positive x-axis, and therefore, $\theta=0$. The distance from the origin is 2. So

$$(r, \theta, z) = (2, 0, -1).$$

The point (-1, 0, 0) lies on the negative x-axis and therefore $\theta=\pi$. The distance from the origin is 1. Therefore,

$$(r, \theta, z) = (1, \pi, 0).$$

The projection of $(\frac{1}{2}, -\frac{\sqrt{3}}{2}, -2)$ onto the xy-plane is $(\frac{1}{2}, -\frac{\sqrt{3}}{2}, 0)$, giving $\tan(\theta) = -\sqrt{3}$. Since this point lies in the quadrant $x > 0$, $y < 0$, we take θ as $\arctan(-\sqrt{3}) = -\frac{\pi}{3}$. We also have $r = \sqrt{x^2 + y^2}$, giving $r = 1$. Therefore,

$$(r, \theta, z) = (1, -\tfrac{\pi}{3}, -2).$$

For $(-\sqrt{2}, \sqrt{2}, 1)$, $\tan(\theta) = -1$. Since $\arctan(-1) = -\frac{\pi}{4}$ and the point lies in the quadrant $x < 0$, $y > 0$, add π to give $\theta = \frac{3\pi}{4}$. We also have $r = \sqrt{x^2 + y^2}$, giving $r = 2$. Therefore,

$$(r, \theta, z) = (2, \tfrac{3\pi}{4}, 1).$$

Exercise 14.4

Cartesians and cylindrical polars

Use the software to explore the relation between Cartesians and cylindrical polars. In particular, verify the conversions between the systems shown in Examples 14.7 and 14.8.

This system is called cylindrical polars because if r is held fixed, then the point P is constrained to move on the surface of a cylinder as θ and z vary. Run the software animation that shows this.

Spherical Polar Coordinates

In three dimensional space, the position of a point P relative to an origin O and Cartesian axes OX, OY, OZ may be specified by *spherical coordinates* (ρ, θ, ϕ). The distance of P from the origin is the length ρ=OP, and ϕ is the angle ZOP between the z-axis and the line OP (see Figure 14.8). Let P′ be the projection of P onto the *xy*-plane; then θ is the angle XOP′ between the *x*-axis and the line OP′, measured positive in the anticlockwise direction in the *x-y* plane.

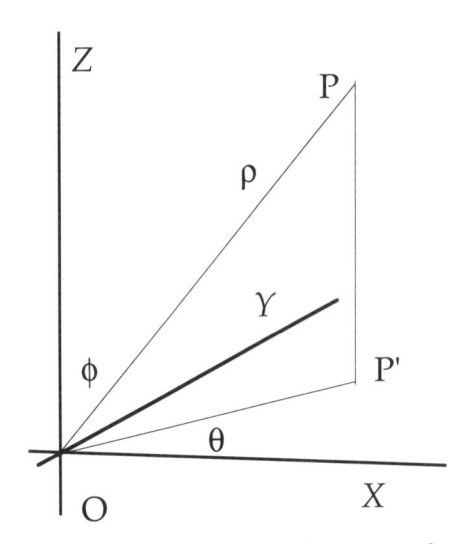

Figure 14.8. Spherical polar coordinates (r, q, f).

We should note the differences between spherical polar coordinates and cylindrical polars. Only the angle θ is common between the two systems, and the cylindrical coordinate *r* does *not* give the distance of P from the origin, whereas in spherical polar coordinates, ρ=OP is used and *does* give the distance of P from the origin.

Observe that from the geometry of the definition, ϕ takes a value in the interval $[0, \pi]$, and the value of θ may be taken in the interval $[-\pi, \pi]$ (although other choices are possible, such as $[0, 2\pi]$).

We can transform spherical polar coordinates to Cartesians by noting first that the *z* coordinate of P is the projection of OP on the *z*-axis, and therefore $z = \rho\cos(\phi)$. Next note that the length *r* of OP′ is the projection of OP on the XY-plane, and therefore $r = \rho\sin(\phi)$, which is the same *r* as used in plane polars. Therefore, we have

$$\left. \begin{array}{l} x = \rho\sin(\phi)\cos(\theta) \\ y = \rho\sin(\phi)\sin(\theta) \\ z = \rho\cos(\phi) \end{array} \right\} \quad (14.6)$$

To convert from Cartesians to spherical polar coordinates, first note that the length OP = $\rho = \sqrt{x^2 + y^2 + z^2}$. We can find ϕ from the last equation in (14.6), which gives $\cos(\phi) = \frac{z}{\rho}$. From the first two equations of (14.6), or direct from the diagram, $\tan(\theta) = \frac{y}{x}$. Therefore,

$$\left. \begin{array}{l} \rho = \sqrt{x^2 + y^2 + z^2} \\ \theta = \arctan\left(\frac{y}{x}\right) + n\pi \\ \phi = \arccos\left(\frac{z}{\rho}\right) \end{array} \right\} \quad (14.7)$$

where *n* is an integer that must be chosen in the same way as in equation (14.2) for plane polar coordinates. The arccos function must be interpreted to give values of ϕ in the interval $[0, \pi]$.

Example 14.9
Transform the following points from spherical polar coordinates to Cartesians:

(ρ, θ, ϕ) = (1, 0, 0), (2, 0, $\frac{\pi}{2}$), (1, $\frac{\pi}{2}$, $\frac{\pi}{2}$), (2 -$\frac{\pi}{4}$, $\frac{\pi}{3}$,), (2, -$\frac{2\pi}{3}$, $\frac{2\pi}{3}$).

For (ρ, θ, ϕ) = (1, 0, 0), observe that $\phi=0$ so the vector OP is aligned with the z-axis. So the corresponding $(x, y, z) = (0, 0, 1)$.

For (ρ, θ, ϕ) = (2, 0, $\frac{\pi}{2}$), observe that $\phi=\frac{\pi}{2}$ so the vector OP lies in the XY-plane. Since $\theta=0$, the vector OP is aligned with the x-axis. So the corresponding $(x, y, z) = (2, 0, 0)$.

For (ρ, θ, ϕ) = (1, $\frac{\pi}{2}$, $\frac{\pi}{2}$), observe that $\phi=\frac{\pi}{2}$ so the vector OP lies in the XY-plane. Since $\theta=\frac{\pi}{2}$ the vector OP is aligned with the y-axis. So the corresponding $(x, y, z) = (0, 1, 0)$.

For (ρ, θ, ϕ) = (2, -$\frac{\pi}{4}$, $\frac{\pi}{3}$), start by noting $\phi=\frac{\pi}{3}$ so that $\cos(\phi)=\frac{1}{2}$, $\sin(\phi)=\frac{\sqrt{3}}{2}$. From (14.6), we have $z = 1$. Now use $\theta=-\frac{\pi}{4}$ so $\cos(\theta)=\frac{1}{\sqrt{2}}$, $\sin(\theta)=-\frac{1}{\sqrt{2}}$. From (14.6) we get $(x, y, z) = (\sqrt{\frac{3}{2}}, -\sqrt{\frac{3}{2}}, 1)$.

For (ρ, θ, ϕ) = (2, -$\frac{2\pi}{3}$, $\frac{2\pi}{3}$), start by noting $\phi=\frac{2\pi}{3}$ so that $\cos(\phi)=-\frac{1}{2}$, $\sin(\phi)=\frac{\sqrt{3}}{2}$. From (14.6), we have $z = -1$. Now use $\theta=-\frac{2\pi}{3}$ so $\cos(\theta)=-\frac{1}{2}$, $\sin(\theta)=-\frac{\sqrt{3}}{2}$. From (14.6) we get $(x, y, z) = (-\frac{\sqrt{3}}{2}, -\frac{3}{2}, -1)$.

Example 14.10
Transform the following points from Cartesian coordinates to spherical polars, expressing θ in the interval $-\pi < \theta \leq \pi$ and ϕ in the interval $0 < \theta \leq \pi$:

(x, y, z)= (0, 2, -1), (-1, 0, 0), ($\frac{1}{2}$, -$\frac{\sqrt{3}}{2}$, -2), (-$\frac{1}{\sqrt{2}}$, $\frac{1}{\sqrt{2}}$, 1).

For $(x, y, z) = (0, 2, -1)$, from (14.7) we have $\rho = \sqrt{5}$. Equation (14.7) also gives $\phi = \arccos\left(-\frac{1}{\sqrt{5}}\right)$ which is an angle greater than $\frac{\pi}{2}$, because the point lies below the xy-plane (z<0). In the xy-plane itself, the projected direction is along the y-axis, and therefore $\theta=\frac{\pi}{2}$. Therefore,

$$(\rho, \theta, \phi) = \left(\sqrt{5}, \frac{\pi}{2}, \arccos\left(-\frac{1}{\sqrt{5}}\right)\right).$$

For $(x, y, z) = (-1, 0, 0)$ we have $\rho = 1$. The direction is along the negative x-axis; therefore, $\phi = \frac{\pi}{2}$ and $\theta=\pi$. We have

$$(\rho, \theta, \phi) = (1, \pi, \tfrac{\pi}{2}).$$

For $(x, y, z) = (\frac{1}{2}, -\frac{\sqrt{3}}{2}, -2)$,

$$\rho^2 = \left(\tfrac{1}{2}\right)^2 + \left(-\tfrac{\sqrt{3}}{2}\right)^2 + (-2)^2 = 5.$$

From (14.7) $\phi = \arccos\left(-\frac{2}{\sqrt{5}}\right)$, so $\phi > \frac{\pi}{2}$. Since $x > 0$ we know $-\frac{\pi}{2} \leq \theta \leq \frac{\pi}{2}$, and therefore $\arctan\left(-\sqrt{3}\right)$ gives $\theta = -\frac{\pi}{3}$. We have

$$(\rho, \theta, \phi) = \left(\sqrt{5}, -\frac{\pi}{3}, \arccos\left(-\tfrac{2}{\sqrt{5}}\right)\right).$$

For $(x, y, z) = (-\frac{1}{\sqrt{2}}, \frac{1}{\sqrt{2}}, 1)$, $\rho=\sqrt{2}$. From (14.7), $\phi = \arccos\left(\frac{1}{\sqrt{2}}\right)$, so $\phi = \frac{\pi}{4}$. Since $x < 0$ and $y > 0$, $\arctan(-1)$ gives $\theta = \frac{3\pi}{4}$. We have

$$(\rho, \theta, \phi) = (\sqrt{2}, \tfrac{3\pi}{4}, \tfrac{\pi}{4}).$$

Exercise 14.5

Cartesians and spherical polars

Use the software to explore the relation between Cartesians and spherical polars. In particular, verify the conversions between the systems shown in Examples 14.1 and 14.2.

This system is called spherical polars because if ρ is held fixed, then the point P is constrained to move on the surface of a sphere as ϕ and θ vary. Run the software animation that shows this.

Exercise 14.6

Converting Cartesians to spherical polar coordinates

Writing the length $OP' = r = \sqrt{x^2 + y^2}$, show that equations (14.6) may be written equivalently as

$$\begin{aligned} \rho &= \sqrt{r^2 + z^2} \\ \theta &= 2m\pi \pm \arccos\left(\frac{x}{r}\right) \\ \phi &= \arctan\left(\frac{r}{z}\right) + n\pi \end{aligned}$$

where the integers m and n and the $\pm$ sign in the expression for θ are chosen to ensure that the direction corresponding to θ is the same as the direction of the vector $(x, y, 0)$, and that ϕ lies in the interval $[0, \pi]$.

Explain how to choose m for every combination of values of x and y, assuming $r \neq 0$. Explain how to choose n for every value of z, assuming $\rho \neq 0$. What happens if $r = 0$?

WORKED EXAMPLE 14.1 TRANSFORMING EQUATIONS

In Cartesian coordinates the equation of an ellipse (see Figure 14.9) with semi-axes of length a, b (where $a > b$) and center at the origin is

$$\frac{x^2}{a^2} + \frac{y^2}{b^2} = 1 \qquad (14.8)$$

This can be put in polar coordinate form with the origin at the center, using the substitutions in equation (14.1):

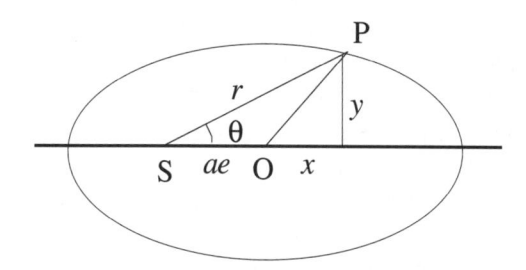

$$x = r\cos(\theta), \quad y = r\sin(\theta).$$

Figure 14.9. Polar coordinates for an ellipse.

Then (14.8) can be rewritten as

$$r^2 = \frac{a^2 b^2}{b^2 \cos^2(\theta) + a^2 \sin^2(\theta)}.$$

Since r is defined to be positive, the positive square root may be taken without ambiguity to give

$$r = ab\left(b^2 \cos^2(\theta) + a^2 \sin^2(\theta)\right)^{-\frac{1}{2}} \qquad (14.9)$$

This is a perfectly good polar coordinate form for an ellipse, but it turns out that a neater form can be found. In the coordinate system used in (14.8), there are two special points ($\pm ae$, 0), called the *foci* of the ellipse, where e is called the eccentricity and is defined by

$$e = \sqrt{1 - \frac{b^2}{a^2}} \qquad (14.10)$$

Now take the origin of polar coordinates at the focus $S = (-ae, 0)$. The coordinates x and y as used in (14.1) are then given by

$$x = r\cos(\theta) - ae, \quad y = r\sin(\theta) \qquad (14.11)$$

Multiply (14.8) by b^2, substitute from (14.11), and use $\sin^2(\theta) = 1 - \cos^2(\theta)$ to get

$$\frac{b^2}{a^2}(r\cos(\theta) - ae)^2 + r^2\left(1 - \cos^2(\theta)\right) = b^2$$

which can be expanded and rearranged as

$$r^2 \cos^2(\theta)\left[\frac{b^2}{a^2} - 1\right] - 2e\frac{b^2}{a}r\cos(\theta)$$
$$-b^2\left(1 - e^2\right) = -r^2.$$

From the definition of e we have

$$\left[\frac{b^2}{a^2} - 1\right] = -e^2 \text{ and } 1 - e^2 = \frac{b^2}{a^2}; \text{ therefore,}$$

$$-\left(re\cos(\theta) + \frac{b^2}{a}\right)^2 = -r^2.$$

Change the signs on both sides; then the square root can be taken and terms rearranged to give finally

$$r(1 - e\cos(\theta)) = \frac{b^2}{a} \qquad (14.12)$$

Note that since $a > b$ (and both are taken as positive) we must also have $0 < e < 1$. The result (14.12) also holds for $e = 0$ (implying $a = b$), which just gives $r = a$, the circle of radius a, where both foci coincide with the center of the circle. Note also that as $|\cos(\theta)| \leq 1$, the term $(1 - e\cos(\theta))$ is never zero, and so r is defined for all values of θ.

As $\frac{b}{a}$ tends to zero, e tends to 1 and the ellipse takes on a more elongated shape. You might like to verify this using the software.

WORKED EXAMPLE 14.2
PROPERTIES OF AN ELLIPSE

Equation (14.12) is a simpler form than (14.9), and it will now be used to prove an important property of the ellipse.

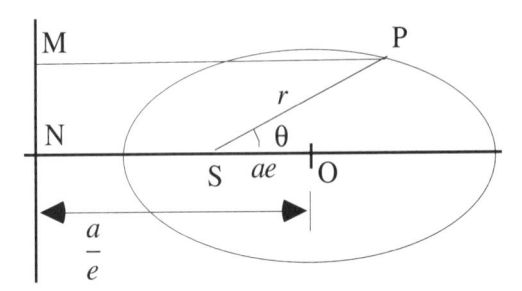

Figure 14.10. Directrix-focus property of an ellipse.

Work with the same ellipse as shown in Worked Example 14.1. S is the focus distance ae to the left of the ellipse's center O, P is any point on the ellipse, and S is used as the origin of polar coordinates (r, θ).

Construct a line, called a *directrix*, at right angles to the direction of the major axis and at a distance $\frac{a}{e}$ from the center O, on the same side of O as S (see Figure 14.10). Let N be the point where this line and the direction of the major axis intersect, and let M be the point which is the projection of P on the directrix, so that PM is parallel to the major axis. Since (as in the previous Worked Example) we know $0 < e < 1$, the distance ON is greater than the distance OS. Using equation (14.10) we find that the distance SN is

$$\text{SN} = \frac{a}{e} - ae = \frac{a}{e}(1 - e^2) = \frac{b^2}{ae}.$$

The distance PM is the sum of SN and the projection of the length SP along the major axis, that is:

$$\text{PM} = \frac{b^2}{ae} + r\cos(\theta).$$

But from (14.12) we have

$$r = \frac{b^2}{a} + er\cos(\theta).$$

Therefore,

$$\text{SP} = e\,\text{PM} \qquad (14.13)$$

This property is often used to define members of the class of curves known as *conics*. In the software, the glossary entry for ellipse starts from equation (14.13) and derives (14.12). When working this way round, there is no need to restrict the value of e (except that it must be positive as we are comparing two lengths). However, if $e > 1$, then the term $(1 - e\cos(\theta))$ in (14.12) can become zero (or negative), and the value of r is no longer defined for all θ. It is also clear that we must reinterpret the meaning of b, since (14.10) will not allow $e > 1$. In fact, the curve defined by $\text{SP} = e\,\text{PM}$ for $e > 1$ is a *hyperbola*. You can see the derivation and further explanation in the glossary entry in the software.

PROBLEM 14.1
POLAR COORDINATES

Name: _____
Date: _____
Section: _____

(i) Complete the table so that each of the following points is given in both Cartesian and plane polar coordinates. Values for θ should be in the range $-\pi < \theta \leq \pi$.

	(x, y)	(r, θ)
A	(2, 1)	(......, ...)
B	(......, ...)	$(\sqrt{6}, \frac{\pi}{6})$
C	(-1, -2)	(......, ...)
D	(......, ...)	$(\sqrt{3}, -\frac{5\pi}{6})$

	(x, y)	(r, θ)
E	(4, -4)	(......, ...)
F	(......, ...)	$(2, \frac{3\pi}{4})$
G	(-3, ...)	(5, ...)
H	(1, ...)	$(......, -\frac{\pi}{3})$

Two answers are possible for point G. Give both.

Which of all these points is closest to the origin? .... distance = ...

Which of all these points is farthest from the origin? .... distance = ...

(ii) Complete the table so that each of the following points is given in Cartesian, cylindrical, and spherical polar coordinates. Answers should be given to two places of decimals. Values for θ should be in the range $-\pi < \theta \leq \pi$, and values for ϕ should be in the range $0 \leq \phi \leq \pi$.

	(x, y, z)	(r, θ, z)	(ρ, θ, φ)
P	(2, -1, 1)	(......, ..., ...)	(......, ..., ...)
Q	(......, ..., ...)	$(\sqrt{6}, 2\pi/3, 1)$	(......, ..., ...)
R	(......, ..., ...)	(......, ..., ...)	$(\sqrt{6}, \pi, 2\pi/3)$
S	(4, ..., ...)	(5, ..., ...)	(5, ..., ...)
T	(......, 3, ...)	(......, $\pi/2$, ...)	(5, ..., ...)
U	(1, ..., ...)	(......, ..., ...)	(2, $\pi/4$, ...)
V	(......, ..., 4)	(......, ..., ...)	(......, ..., 0)
W	(......, ..., ...)	$(2\sqrt{2}, -3\pi/4, ...)$	(......, ..., $3\pi/4$)

For which points are two or more answers possible? List the alternatives.

[Continued overleaf...]

PROBLEM 14.1
POLAR COORDINATES

(Continued)

(ii) continued...

Which of the points P-W is closest to the origin?   ....

Which of the points P-W is farthest from the origin? ....

(iii) Find the distances between the following pairs of points as defined in question (i). Give your answers to two places of decimals.

AB = ...   AD = ...   AF = ...

BC = ...   BE = ...   BH = ...

If G1 and G2 are the two distinct answers for G,
what is the distance between G1 and G2?   ....

(iv) Find the distances between the following pairs of points used in questions (i) and (ii). The points A-H in question (i) should be regarded as lying in the plane $z = 0$. Give your answers to two places of decimals.

AP = ...   BQ = ...   CR = ...

PV = ...   RV = ...   VW = ...

Give the distances between the distinct alternatives for the answers to point U.

How many distinct answers can be found for point S?

    Give the distance between the two answers with $z > 0$:   ....

    Give the distance between the two answers with $y > 0$:   ....

PROBLEM 14.2
POLAR COORDINATES

Name: _____

Date: _____

Section: _____

(i) Using cylindrical coordinates (r, θ, z) in 3-D space, describe the surfaces:

r = constant . . .

θ = constant . . .

z = constant . . .

(ii) Using spherical polar coordinates (ρ, θ, ϕ) in 3-D space, describe the surfaces:

ρ = constant . . .

θ = constant . . .

ϕ = constant . . .

(iii) Show that in plane polar coordinates (r, θ), the Cartesian equation for a line in the (x, y) plane $y = mx + c$ may be written

$$r\sin(\theta - \alpha) = \frac{c}{\sqrt{1+m^2}}.$$

[Hint: An expression of form $a\cos(\theta) + b\sin(\theta)$ may always be expressed in the form $R(\sin(\alpha)\cos(\theta) + \cos(\alpha)\sin(\theta))$ where $R = \sqrt{a^2 + b^2}$ and $\tan(\alpha) = \frac{a}{b}$.]

Hence find the slope and intercept with the y-axis of the line $r\sin(\theta - \pi/4) = \frac{1}{\sqrt{2}}$.

Slope m = ...   Intercept c = ...

Express this line in the form $r\cos(\theta + \beta) = d$.

β = ... d = ...

PROBLEM 14.2
POLAR COORDINATES
(Continued)

(iv) Apply the method of Worked Example 14.2 to a hyperbola.

Start with the equation of a hyperbola $\dfrac{x^2}{a^2} - \dfrac{y^2}{b^2} = 1$, and define $e = \sqrt{1 + \dfrac{b^2}{a^2}}$.

Show that if the point $S = (ae, 0)$ (relative to Cartesian coordinates with origin at the center of the hyperbola) is taken as the origin for polar coordinates, then the equation of the hyperbola may be written $r(e\cos(\theta) - 1) = d$, and find d.

$$d = \ldots\ldots$$

Sketch one branch of the hyperbola, showing the x-axis, and point S.

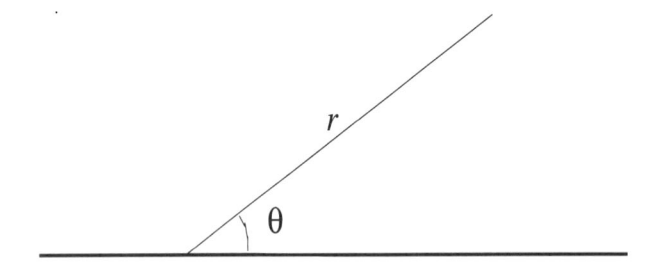

Show that there is an angle α such that $r \to \infty$ as $\theta \to \pm\alpha$. Show that $\tan(\alpha) = \dfrac{b}{a}$.

Find and sketch a directrix, analogous to the directrix found for the ellipse in Worked Example 14.2, and prove the property $SP = e\,PM$ for the hyperbola.

NEW SITUATIONS

1. Areas using polars

In Cartesian coordinates areas under curves described by equations $y = f(x)$ were found by evaluating definite integrals like

$$\int_a^b f(x)dx.$$

Here we show that equivalent techniques can be used for finding areas enclosed by curves of the form $r = f(\theta)$, where (r, θ) are plane polar coordinates.

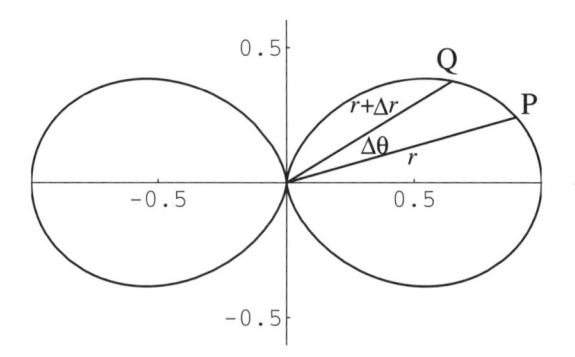

Figure 14.11. The curve $r = \cos^2(\theta)$.

Instead of dividing the area into strips (as was done for Cartesian curves), we use infinitesimal triangles. In Figure 14.11, P is the point (r, θ) and Q is a point $(r+\Delta r, \theta+\Delta\theta)$. Since $\Delta\theta$ is small, the area of triangle OPQ is approximately $\frac{1}{2}r^2\Delta\theta$. As $\Delta\theta \to 0$ this becomes more accurate, and in the limit, summing all these areas, we have

$$\text{area} = \int_\alpha^\beta \tfrac{1}{2}r^2 d\theta \qquad (14.14)$$

where $[\alpha, \beta]$ is the interval corresponding to the values of θ traversed by the point P. The area obtained is the area enclosed by the radius vectors OP at these limits and the segment of the curve traversed.

As a simple check, consider the circle $r = a$ over the interval $-\pi \leq \theta \leq \pi$, giving

$$\text{area} = \int_{-\pi}^{\pi} \tfrac{1}{2}a^2 d\theta = \pi a^2.$$

For the curve $r = \cos^2(\theta)$ illustrated in Figure 14.11, the limits for the loop on the right-hand side are $\alpha = -\frac{\pi}{2}$ and $\beta = \frac{\pi}{2}$. The area of the right-hand loop is therefore

$$\int_{-\pi/2}^{\pi/2} \tfrac{1}{2}\cos^4(\theta)d\theta.$$

Integrals like these may be evaluated using a variety of techniques; see, for example, the reduction formula method suggested in Problems 9.2 (Chapter 9). In this case the result is $\frac{3\pi}{16}$.

2. Arc length in polars

In Chapter 9, example (vi) of Worked Example 9.2, the arc length of a curve $y = f(x)$ from $x=a$ to $x=b$ was shown to be

$$L(a,b) = \int_a^b \sqrt{1+\left(\frac{dy}{dx}\right)^2}\, dx.$$

To obtain an analogous formula for use with curves of the form $r = f(\theta)$, consider the diagram in Figure 14.12.

As in Figure 14.11, P and Q are two points on the curve corresponding to values θ and $\theta+\Delta\theta$. R is the point on OQ distance r from O. For small $\Delta\theta$, the line PR has length approximately $r\Delta\theta$. PQR is approximately a right-angled triangle, and the arc length PQ along the curve is therefore

$$\Delta s \approx \sqrt{r^2\Delta\theta^2 + \Delta r^2}.$$

In the limit as $\Delta\theta \to 0$ the errors in the approximations approach zero giving

$$\Delta s \approx \sqrt{r^2 + \left(\frac{dr}{d\theta}\right)^2} \Delta\theta$$

and hence arc length is given by

$$L = \int_\alpha^\beta \sqrt{r^2 + \left(\frac{dr}{d\theta}\right)^2} \, d\theta \qquad (14.15)$$

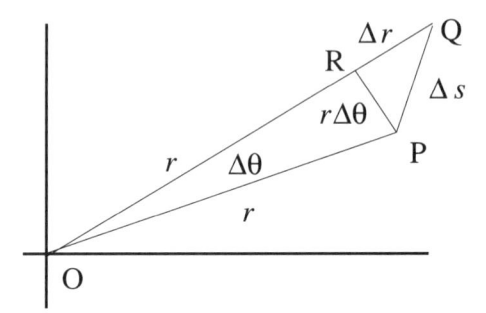

Figure 14.12. A segment of arc PQ.

For the curve $r = \cos^2(\theta)$, we have $r' = -2\cos(\theta)\sin(\theta)$ and hence

$$L = \int_\alpha^\beta \cos(\theta)\left(\cos^2(\theta) + 4\sin^2(\theta)\right)^{1/2} d\theta.$$

No closed form can be found for this integral.

Chapter 15

Parametric Equations

Prerequisites

Before you study this material, you should be familiar with:

(1) Geometrical idea of lines and equations (Module 1) and basic concepts of the Calculus. (Modules 3 and 4).

(2) The solution of differential equations (Module 13).

(3) Vectors, in particular unit vectors in the direction of the coordinate axes.

Objectives

In this chapter we introduce the idea of parameterizing curves. By now you should be familiar with the notion that if we are given a function f, then plotting the points $(x, f(x))$ will generate a curve. Each point on the curve is determined by the value given to x; effectively the value of x parameterizes the curve. In this chapter we extend this idea to other coordinate systems and other parameterizations.

This module also shows how the Calculus can be used to provide a theoretical basis for the mathematical modeling of moving objects. Although many simplifying assumptions are made, the basic principles carry through into more advanced applications.

Connections

When trying to describe the motion of moving objects, time itself is a natural parameterization. As a simple example, a baseball after being struck will follow a curve that could be plotted as a graph. We expect the laws of nature to tell us something about the horizontal and vertical speeds, and hence how each of the coordinates x and y of the baseball change with time. In mathematical terms, we are expecting to find $(x(t), y(t))$. Once these are known, it might be possible to "eliminate t", and write y as a function of x. Any way of expressing a curve $y(x)$ in the form $(x(t), y(t))$ is called a parameterization. However, the parameter t is not necessarily time. This will be illustrated for a selection of standard plane curves, such as lines and conics (ellipses, circles, parabolas and, hyperbolas).

The videos are designed to motivate the use of other parameterizations as well as time.

The first application examines the motion of a ski jumper to illustrate how different parameterizations may be appropriate at different times. For example, when a jumper is moving down the ramp, prior to takeoff, his motion is determined by the shape of the ramp. We can fix his position by his horizontal displacement and then find his exact location from the shape function corresponding to the ramp. Alternatively, we can simply measure the distance he has traveled along the ramp itself. Or we can measure the time elapsed since he started the run. Any of these methods would enable us to parameterize the position of the jumper in terms of a single variable.

Once the skier leaves the end of the ramp and is in flight, we need three variables to specify his position, $x(t)$, $y(t)$, and $z(t)$. In a simple situation where there is no cross wind, we can assume that he continues in the plane of the ramp, in which case we need only two variables, horizontal and vertical displacements. Newton's Laws of Motion can then be used to derive expressions for these functions.

The second application looks at motion on a much larger scale: the motion of the planets in the solar system. This application explains why using polar coordinates (r,θ) is a more natural choice of coordinates than Cartesian coordinates (x,y), and why taking θ to determine the parameterization leads to significant simplification of the equations that result. We will show that, although the background theory of the motion of the planets is beyond the scope of the module, only an elementary knowledge of polar coordinates gives sufficient information to investigate the motion of the planets.

In this module we look more closely at the idea of parameterizing paths and how we can use the information gained to study motion. This is where the origins of the Calculus lie and give ample demonstration of its power.

Polar Coordinates Review

The Cartesian coordinates of a point, (x,y), and its polar coordinates, (r,θ), are related by

$$x = r\cos(\theta), \quad y = r\sin(\theta)$$

as shown in Figure 15.1.

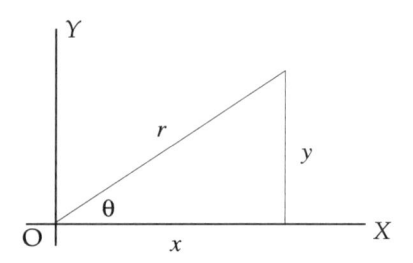

Figure 15.1. Cartesian and polar coordinates.

Alternatively, we can write

$$r^2 = x^2 + y^2, \qquad \tan(\theta) = \frac{y}{x},$$

which enables us to find r and θ provided x and y are known and we take care as to the quadrant in which the point (x,y) lies.

In Cartesian coordinates we plot $y = f(x)$ by substituting a number of values for x, determine y, and then plot the points, (x,y), obtained in this way. We can think of the curve $y = f(x)$ being parameterized by the value of x.

In polar coordinates (r, θ), a curve can be specified by $r = f(\theta)$; that is, once θ is given we can determine r. We can think of the curve $r = f(\theta)$ being parameterized by the value of θ. Recall that since θ is the polar angle we need only consider either $0 \le \theta < 2\pi$ or $-\pi < \theta \le \pi$.

Example 15.1

The curve $r = 1 + 2\cos(2\theta)$ is shown in Figure 15.2. To get this curve we tabulate r as a function of θ. (See Table 15.1 below.) Plotting these points and using symmetry gives the curve as shown.

θ	0	$\frac{\pi}{6}$	$\frac{\pi}{4}$	$\frac{\pi}{3}$	$\frac{\pi}{2}$
r	3	2	1	0	-1

Table 15.1. $r = 1 + 2\cos(2\theta)$ for $0 \le \theta \le \frac{\pi}{2}$.

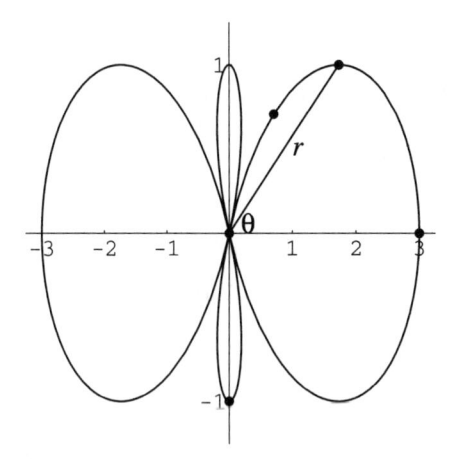

Figure 15.2. The curve $r = 1 + 2\cos(2\theta)$, $0 \le \theta \le 2\pi$.

Exercise 15.1

Plot the following curves:

(a) $r = 1 - 2\cos(3\theta)$, $0 \le \theta < 2\pi$,
(b) $r = 1 - 2\sin(4\theta)$, $-\pi < \theta \le \pi$.

Exercise 15.2

Check the results of Exercise 15.1 using the plotter provided with *Calculus Connections*.

General Parametric Coordinates x=x(t), y=y(t)

The curve $y = x^2$ is a familiar curve that can be plotted since for each value of x we can associate a single value for y. For example, when $x=2$, then $y=4$, and we plot the point (2,4). As the value of x changes, the value of y changes in a pre-determined way; we say that y is a function of x. The variable y is said to be the dependent variable and x the independent variable since it changes freely.

However, if x does not change freely but only according to some expression $x=x(t)$, then y will also depend on t so that $y=y(t)$. The value of t parameterizes the curve $(x(t), y(t))$.

Example 15.2

The curve $x = r\cos(t)$, $y = r\sin(t)$ represents the equation of the circle shown in Figure 15.3 since

$$x^2 + y^2 = r^2(\cos^2(t) + \sin^2(t)) = r^2.$$

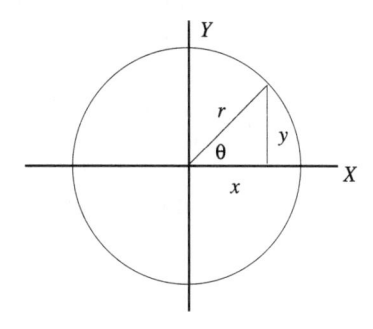

Figure 15.3. Parameterization of the circle in Example 15.2.

Later, we will see that selecting an appropriate parameterization can be very helpful in simplifying problems.

Examples: Conic Sections

As examples of parameterizations, we will look at the idea of conic sections. If an inclined line, which passes through a fixed point. is rotated, it will draw out a cone as shown in Figure 15.4.

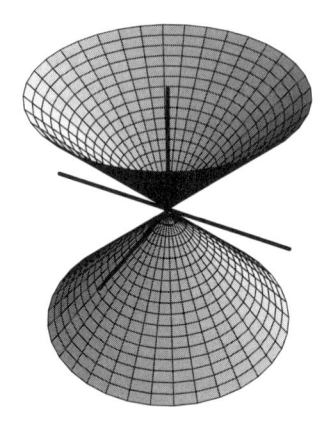

Figure 15.4. Cones generated by an inclined line.

Now imagine a plane that cuts through the cones. The way in which the cones are cut will produce a different cross section; the result is called a *conic section*.

A horizontal cut through a cone will produce a circle as shown in Figure 15.5. If we assume that the circle is centered at the origin, then the equation of the circle is

$$x^2 + y^2 = r^2.$$

However, we can parameterize the circle by taking

$$x = r\cos(\theta), \ y = r\sin(\theta), \ 0 \le \theta < 2\pi,$$

as was shown in Example 15.2.

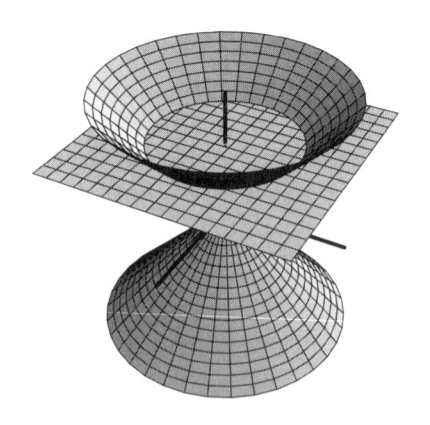

Figure 15.5. A horizontal plane intersecting a cone to give a circle.

When we use an inclined plane, we get a cross section in the shape of an ellipse that has the equation

$$\frac{x^2}{a^2} + \frac{y^2}{b^2} = 1$$

which we can parameterize as

$$x = a\cos(\phi), \ y = b\sin(\phi), \ 0 \le \phi < 2\pi.$$

As the angle of inclination of the plane increases, until it is parallel with one of the lines that generates the cones, the cut will give a parabola with the Cartesian equation

$$y^2 = 4ax$$

which can be parameterized by

$$x = at^2, \ y = 2at, \ 0 \le t < \infty.$$

Ultimately, the plane will cut both cones and we will get the two branches of a hyperbola in a vertical plane XY, (see Figure 15.6), that is,

$$\frac{x^2}{a^2} - \frac{y^2}{b^2} = 1.$$

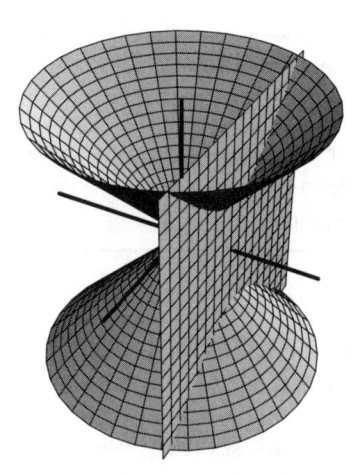

Figure 15.6. A vertical plane cutting both cones to give the two branches of a hyperbola.

This hyperbola can be parameterized as

$x = a\sinh(\theta)$, $y = \pm b\cosh(\theta)$, $-\infty \leq \theta < \infty$.

Circles, parabolas, hyperbolas, and ellipses are all examples of the general conic section or curves that can be written in the form

$$Ax^2 + Bxy + Cy^2 + Dx + Ey + F = 0.$$

If $B=0$, and A and C have the same sign then this equation will generate an ellipse. If A and C have the opposite sign then we will get a hyperbola. If either $A=C=0$ or $B=C=0$ then we get a parabola. In general, the type of conic section generated depends on the sign of B^2-AC. Each different type requires a different parameterization.

Exercise 15.3

Conic sections

Determine whether the following conic sections are ellipses, parabolas, or hyperbolas.

(a) $3x^2 + y^2 = 1$,

(b) $3x^2 - y^2 = 1$,

(c) $3x^2 + 2xy - y^2 = 1$.

In each case suggest a suitable parameterization.

Piecewise Parametric Curves

In Example 15.2 we parameterized a circle using just one set of expressions. Sometimes we need to use different parameterizations on different parts of a curve.

Example 15.3

Consider the path shown in Figure 15.7 which is made up from the quarter circle BC and the straight lines CA and AB.

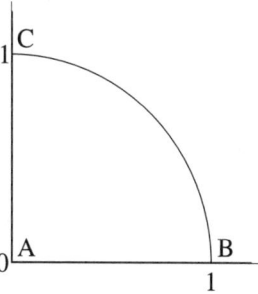

Figure 15.7. A quarter disc as described in Example 15.3.

We can parameterize the quarter circle in Figure 15.7 as follows.

AB: $x(t) = t$, $y(t) = 0$, $0 < t \leq 1$,

BC: $x(t) = \cos(t)$, $y(t) = \sin(t)$, $0 < t \leq \frac{\pi}{2}$,

CA: $x(t) = 0$, $y(t) = 1-t$, $0 \leq t < 1$.

The parameterization of the curve in Figure 15.7 is called a piecewise parameterization.

Velocity and Acceleration in Cartesian Coordinates

In the previous sections, we saw how the position of a particle could be parameterized by time t as $x = x(t)$, $y = y(t)$. This gives the displacement in the x and y directions. Since these are indepen-

118 Calculus Connections

dent of each other, we see that the average change in position in the x direction over an interval Δt is given by

$$\frac{x(t+\Delta t)-x(t)}{\Delta t}.$$

As $\Delta t \to 0$ this becomes $\frac{dx}{dt}$. Likewise, the rate of change in the y direction is $\frac{dy}{dt}$.

Similarly, the acceleration parallel to the x direction is x'', and parallel to the y direction it is y''.

Example 15.4
Find the velocity and acceleration of a particle moving around a circle of radius 1. The position of the particle is given by the parameterization $x = \cos(t)$, $y = \sin(t)$.

$$x(t) = \cos(t), \quad \frac{dx}{dt} = -\sin(t), \quad \frac{d^2 x}{dt^2} = -\cos(t),$$

$$y(t) = \sin(t), \quad \frac{dy}{dt} = \cos(t), \quad \frac{d^2 y}{dt^2} = -\sin(t).$$

Notice that the velocity is perpendicular to the displacement vector $(x(t), y(t))$ but that the acceleration vector points back along the direction of this vector. (See Figure 15.8.) Notice, too, that even though the speed is constant,

$$\left(\frac{dx}{dt}\right)^2 + \left(\frac{dy}{dt}\right)^2 = \sin^2(t) + \cos^2(t) = 1,$$

the particle is accelerating since its velocity is changing direction.

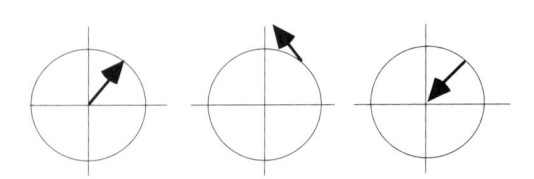

Figure 15.8. The displacement, velocity, and acceleration of the particle in Example 15.4.

Exercise 15.4
Determine the velocity and acceleration of the particle described in the piecewise parameterization in Example 15.2.

What If The Parameter Is Not Time?

In the previous section, we saw how the position of a particle could be parameterized by time t as $x = x(t)$, $y = y(t)$ and how such a parameterization could be used to find the velocity and acceleration of the particle. We can now consider what happens if we use a different parameterization, $x = x(s)$, $y = y(s)$. To find the velocity in the x direction we need $\frac{dx}{dt}$. To do this we use the chain rule to get

$$v_x = \frac{dx}{dt} = \frac{dx}{ds}\frac{ds}{dt}, \quad v_y = \frac{dy}{dt} = \frac{dy}{ds}\frac{ds}{dt}.$$

Similarly, we can find the acceleration parallel to the x and y directions as

$$\frac{dv_x}{dt} = \frac{dv_x}{ds}\frac{ds}{dt}, \quad \frac{dv_y}{dt} = \frac{dv_y}{ds}\frac{ds}{dt}.$$

Exercise 15.5
Find the velocity and acceleration of a particle whose position is given by the parameterization

$$x = \cos(s), \quad y = \sin(s), \quad s = 2\pi t.$$

Finding Position Given Acceleration in Cartesian Form

We have seen that given the position of a particle in the form $x(t)$, $y(t)$, we can determine its velocity in terms of its components in the x and y directions. Similarly, we can find x'' and y''.

Conversely, given the accelerations x'' and y'' in the directions of the x- and y-axes, we can integrate to find the velocities in the x and y direction. Then we integrate again to find the position $x(t)$ and $y(t)$. However, if we are given

$$\frac{d^2x}{dt^2}, \frac{d^2y}{dt^2}$$

then we will need two additional conditions for each to determine $x(t)$ and $y(t)$.

> **Exercise 15.6**
>
> Given that a particle's acceleration is given by
> $$x''(t) = t, \quad y''(t) = 2,$$
> $$x(0) = 1, \quad x'(0) = 2,$$
> $$y(0) = 0, \quad y'(0) = 1.$$
> Determine the position of the particle after two time units.

Velocity and Acceleration in Polar Coordinates

How do we find velocity and acceleration of a particle if its position is given in polar coordinates?

> The most important point to make is that if $r=r(t)$ and $\theta=\theta(t)$ the velocity is NOT given by $r'(t)$ and $\theta'(t)$.

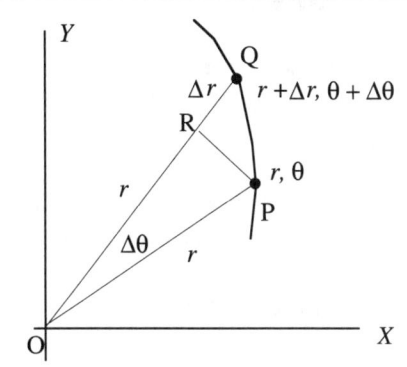

Figure 15.9. Radial and angular speed.

Consider a point moving on a curve from the point P to the point Q, as shown in Figure 15.9, over a time interval Δt. In this interval the radius vector changes by an amount Δr and the angle θ by $\Delta\theta$. The component of velocity in the direction OQ is called the *radial velocity* and its magnitude is given by

Radial speed

$$\frac{QR}{\Delta t} = \frac{\Delta r}{\Delta t} \to \frac{dr}{dt} = r'.$$

Similarly, the component of velocity perpendicular to the direction OQ is called the *transverse velocity*, and its magnitude is given by

Transverse speed

$$\frac{PR}{\Delta t} = \frac{r\Delta\theta}{\Delta t} \to r\frac{d\theta}{dt} = r\theta'.$$

Example 15.5

Find the radial and transverse speed of a particle which moves so that $r=t$, $\theta=2\pi t$.

Radial speed: $\dfrac{dr}{dt} = 1$.

Transverse speed: $r\dfrac{d\theta}{dt} = (t)(2\pi) = 2\pi t$.

Unit Vectors

In Cartesian coordinates the unit vectors in the direction of the *X*- and *Y*-axes are usually denoted by

i in the direction of the *X* axis,
j in the direction of the *Y* axis,

as shown in Figure 15.10.

In polar coordinates the unit vectors are

$\mathbf{e}_r$ is the direction of the radial unit vector **r**
$\mathbf{e}_\theta$ is the direction perpendicular to $\mathbf{e}_r$.

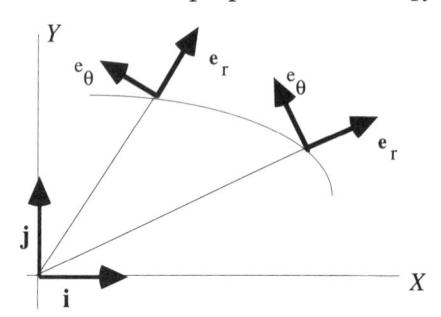

Figure 15.10. The rotation of unit vectors in polar coordinates.

The position of a point in Cartesian coordinates is given by

$$\mathbf{r}(t) = x(t)\,\mathbf{i} + y(t)\,\mathbf{j}$$

but in polars it is simply

$$\mathbf{r}(t) = r(t)\,\mathbf{e}_r$$

Notice that in Cartesian coordinates the unit vectors always point in the direction of the fixed axes. However, in polar coordinates the directions of the unit vectors change as we move along a curve.

Velocity and Acceleration in Cartesian Coordinates

Since the position of any point is given by

$$\mathbf{r}(t) = x(t)\,\mathbf{i} + y(t)\,\mathbf{j}$$

where **i** and **j** are fixed we can determine velocity as

$$\mathbf{r}' = x'\,\mathbf{i} + y'\,\mathbf{j}.$$

Similarly, acceleration is given by

$$\mathbf{r}'' = x''\,\mathbf{i} + y''\,\mathbf{j}$$

as before.

Velocity and Acceleration in Polar Coordinates

From Figure 15.11 we see that the unit vector in the radial direction is given by

$$\mathbf{e}_r = \mathbf{i}\cos\theta + \mathbf{j}\sin\theta.$$

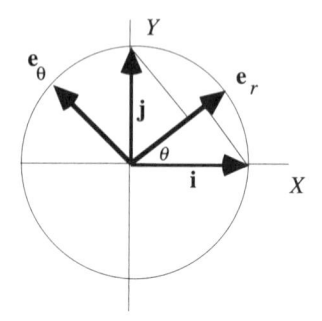

Figure 15.11. Cartesian and polar unit vectors.

Similarly,

$$\mathbf{e}_\theta = -\mathbf{i}\sin\theta + \mathbf{j}\cos\theta.$$

Differentiating these expressions gives

$$\mathbf{e}'_r = -\mathbf{i}\sin\theta\,\theta' + \mathbf{j}\cos\theta\,\theta' = \theta'\,\mathbf{e}_\theta,$$
$$\mathbf{e}'_\theta = -\mathbf{i}\cos\theta\,\theta' - \mathbf{j}\sin\theta\,\theta' = -\theta'\,\mathbf{e}_r.$$

Therefore, if
$$\mathbf{r}(t) = r(t)\mathbf{e}_r$$
then
$$\mathbf{r}'(t) = r'(t)\mathbf{e}_r + r(t)\mathbf{e}'_r = r'(t)\mathbf{e}_r + r(t)\theta'\mathbf{e}_\theta.$$

The radial speed is therefore r' and the transverse speed is $r\theta'$ as before. Similarly,

$$\begin{aligned}\mathbf{r}'' &= \frac{d}{dt}(r'\mathbf{e}_r + r\theta'\mathbf{e}_\theta) \\ &= r''\mathbf{e}_r + r'\mathbf{e}'_r + r'\theta'\mathbf{e}_\theta + r\theta''\mathbf{e}_\theta + r\theta'\mathbf{e}'_\theta \\ &= r''\mathbf{e}_r - (\theta')^2 r\mathbf{e}_r + 2r'\theta'\mathbf{e}_\theta + r\theta''\mathbf{e}_\theta \\ &= (r'' - (\theta')^2 r)\mathbf{e}_r + \frac{1}{r}\frac{d}{dt}(r^2\theta')\mathbf{e}_\theta\end{aligned}$$

The radial component of acceleration in polar coordinates has magnitude

$$\left(r'' - (\theta')^2 r\right)$$

and the transverse component has magnitude

$$\frac{1}{r}\frac{d}{dt}(r^2\theta').$$

Example 15.6

Determine the velocity and acceleration of a particle in uniform motion in a circle,

$$r = R, \ \theta = 2\pi t.$$

Applying the previous results gives

$$\mathbf{r}' = 2\pi R \ \mathbf{e}_\theta.$$

Therefore, the velocity is always in the direction of $\mathbf{e}_\theta$. However,

$$\mathbf{r}'' = -(4\pi^2 R)\mathbf{e}_r,$$

and so the acceleration points toward the origin as in Example 15.4.

Exercise 15.7

The parametric curve

$$r = 1 + 2\cos(2\theta), \ 0 \leq \theta < 2\pi,$$

is plotted in Example 15.1. If a particle completes an orbit in two units of time, determine its velocity, speed, and acceleration at each value of t.

WORKED EXAMPLE 15.1
SKI JUMPER

Let's go back and look again at the ski jumper application. For simplicity, we shall assume that the ramp is an inclined plane given by y=kx, and that the skier, mass m, starts at height H where he pushes off to attain an initial speed of U km per hour parallel to the slope. At the takeoff point, the ramp is angled so that the skier is able to project himself at an angle θ to the horizontal.

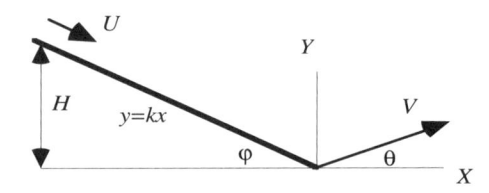

Figure 15.12. The two phases of the ski jump.

Let $v(s)$ be the velocity parallel to the slope at a distance s along the ramp; then from Newton's Laws of Motion,

$$m\frac{dv}{dt} = mg\sin\varphi,$$

which is the component of the acceleration due to gravity parallel to the slope. To integrate this equation, we write this equation as

$$\frac{dv}{ds}\frac{ds}{dt} = v\frac{dv}{ds} = \frac{1}{2}\frac{d}{ds}(v^2) = g\sin\varphi,$$

so that

$$v^2 = 2gs\sin\varphi + C.$$

where C is a constant of integration. However, when s=0 the velocity is v=U and so

$$v^2 = U^2 + 2gs\sin\varphi.$$

The velocity is therefore parameterized by s. At the takeoff point the velocity is given by $v^2 = U^2 + 2gH$, where $\sin\varphi = \frac{H}{S}$.

We shall now assume that the takeoff ramp allows us to consider the subsequent motion as equivalent to a simple projectile that is fired at an angle θ with velocity V. Applying Newton's Laws to the skier with origin at the lip of the ramp, taking x as the horizontal and y as the vertical displacement, we obtain

$$my'' = -mg, \quad mx'' = 0.$$

Integrating these equations gives

$$y' = -gt + C, \text{ and } x' = D.$$

At takeoff, if we reset t=0, so that

$$x'(0) = V\cos\theta, \quad y'(0) = V\sin\theta$$

then

$$x' = V\cos\theta, \quad y' = -gt + V\sin\theta.$$

Integrating again gives

$$y = -\frac{1}{2}gt^2 + Vt\sin\theta, \quad x = Vt\cos\theta.$$

Here the position of the ski jumper is fixed in terms of g, V and θ and the parameter t. We can recover an expression for y in terms of x by solving the expression for x in terms of t to give

$$t = \frac{x}{V\cos\theta}$$

and substituting into y to give

$$y = -\frac{1}{2}g\left(\frac{x}{V\cos\theta}\right)^2 + \frac{\sin\theta}{\cos\theta}x,$$

which is a parabola.

WORKED EXAMPLE 15.2
DERIVATION OF ELLIPTIC ORBIT EQUATIONS

By patient observation, over many years Johannes Kepler recorded the position of the planets in our solar system and deduced that they move in ellipses around the sun and do so in a very precise way. However, he was unable to provide the mathematics behind his conclusions. This was left to Sir Isaac Newton who was able to show, based on very simple assumptions, that the paths of the planets could be determined very precisely.

Newton's inspiration is called the Law of Universal Gravitation. This law states that any two bodies attract each other with a force that is proportional to their masses and inversely proportional to the square of the distance between them. (See Figure 15.13.)

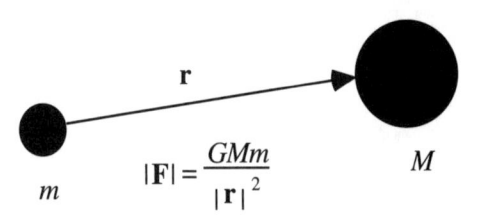

Figure 15.13. The Law of Universal Gravitation.

In addition, Newton's Laws of Motion state that if an object of mass m is traveling so that its displacement is given by the vector $\mathbf{r}$ and it is subject to a force $\mathbf{F}$, then

$$m\mathbf{r}'' = \mathbf{F}.$$

One consequence of the law of Universal Gravitation is that the force $\mathbf{F}$ only has a component in the direction of $\mathbf{r}$ so that the components of acceleration

$$\mathbf{r}'' = \left(r'' - r(\theta')^2\right)\mathbf{e}_r + \frac{1}{r}\frac{d}{dt}\left(r^2\theta'\right)\mathbf{e}_\theta$$

give the radial equation

$$m\left(r'' - r(\theta')^2\right) = -\frac{GMm}{r^2}$$

and the transverse equation

$$m\frac{1}{r}\frac{d}{dt}\left(r^2\theta'\right) = 0.$$

This last equation gives

$$r^2\theta' = C \quad \Rightarrow \quad \theta' = \frac{C}{r^2},$$

where C is an arbitrary constant. Substituting into the radial equation gives

$$r'' - \left(\frac{C}{r^2}\right)^2 r = -\frac{GM}{r^2}$$

which is a second order nonlinear ordinary differential equation for r as a function of the parameter t. We now make a change of variable and parameter and set

$$r(t) = \frac{1}{u(\theta)}$$

so that

$$\frac{dr}{dt} = \frac{d}{dt}\left((u(\theta))^{-1}\right) = \frac{d}{d\theta}\left((u(\theta))^{-1}\right)\frac{d\theta}{dt}$$

$$= -(u(\theta))^{-2}\frac{du}{d\theta}\frac{d\theta}{dt} = -r^2\frac{C}{r^2}\frac{du}{d\theta} = -C\frac{du}{d\theta}$$

and

$$\frac{d^2r}{dt^2} = \frac{d}{dt}\left(-C\frac{du}{d\theta}\right) = \frac{d}{d\theta}\left(-C\frac{du}{d\theta}\right)\frac{d\theta}{dt}$$

$$= -C\frac{d^2u}{d\theta^2}\frac{d\theta}{dt} = -\frac{C^2}{r^2}\frac{d^2u}{d\theta^2}.$$

The nonlinear differential equation
$$r'' - \frac{C^2}{r^3} = -\frac{GM}{r^2}$$
simplifies quite remarkably to
$$\frac{d^2u}{d\theta^2} + u = \frac{GM}{C^2}.$$
This is a linear second order differential equation with constant coefficients, which has the solution
$$u(\theta) = A\cos(\theta + \phi) + B,$$
where A and ϕ are arbitrary constants and
$$B = \frac{GM}{C^2}$$
which is a fixed constant for any planetary system. Finally, we return to the original variable r and parameter t to get
$$r(\theta) = \frac{1}{u(\theta)} = \frac{1}{A\cos(\theta + \phi) + B}.$$
To see that this represents the elliptic orbit which Kepler observed, we write it as
$$r(\theta) = \frac{d}{1 + e\cos(\theta)}.$$
If we plot this function using polar coordinates, then we obtain an ellipse, provided $e<1$. (See Chapter 14.)

PROBLEM 15.1
SIMPLE
PARAMETRIC CURVES

Name: _____
Date: _____
Section: _____

1. Pair the following curves and parametric equations.

   (a)

   [Y-axis graph showing a triangle with vertices at (0,0), (1,0), and (0,1) with hypotenuse]

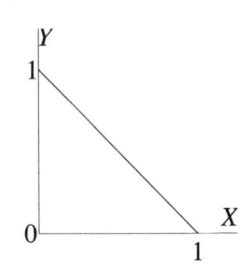

   (b)

   [Graph showing a quarter circle arc from (1,0) to (0,1)]

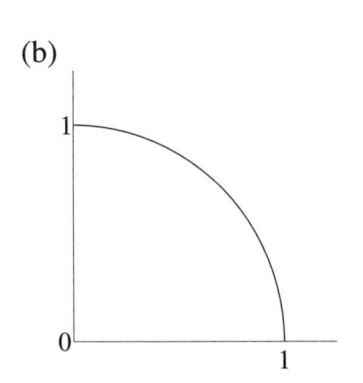

   (c)

   [Graph showing a triangle with vertices at (0,0), (1,0), and (1,1)]

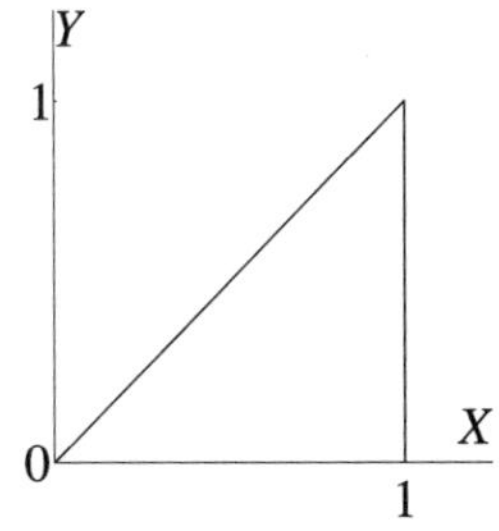

   [ ]
   $x = \cos(\theta), \quad y = \sin(\theta), \quad 0 \le \theta \le \dfrac{\pi}{2},$
   $x = t, \quad y = 0, \quad 0 \le t \le 1,$
   $x = 0, \quad y = t, \quad 0 \le t \le 1,$

   [ ]
   $x = \cos(\theta), \quad y = \sin(\theta), \quad 0 \le \theta \le \dfrac{\pi}{2},$
   $x = t, \quad y = 0, \quad 0 \le t \le 1,$
   $x = 0, \quad y = 1 - t, \quad 0 \le t \le 1,$

   [ ]
   $x = t, \quad y = 0, \quad 0 \le t \le 1,$
   $x = 1 - t, \quad y = 0, \quad 0 \le t \le 1,$
   $x = 0, \quad y = t, \quad 0 \le t \le 1.$

   [ ]
   $x = t, \quad y = t, \quad 0 \le t \le 1,$
   $x = 1, \quad y = t, \quad 0 \le t \le 1,$
   $x = t, \quad y = 0, \quad 0 \le t \le 1.$

   [ ]
   $x = t, \quad y = t, \quad 0 \le t \le 1,$
   $x = 1, \quad y = t, \quad 0 \le t \le 1,$
   $x = t, \quad y = 0, \quad -1 \le t \le 1.$

   [ ]
   $x = t^2, \quad y = t^2, \quad 0 \le t \le 1,$
   $x = 1, \quad y = t, \quad 0 \le t \le 1,$
   $x = t, \quad y = 1 - t^2, \quad -1 \le t \le 1.$

126 Calculus Connections

PROBLEM 15.1
SIMPLE PARAMETRIC CURVES
(Continued)

2. Specify suitable parameterizations of the following curves taken in the order ABC.

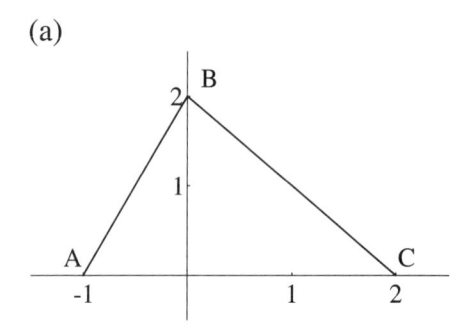

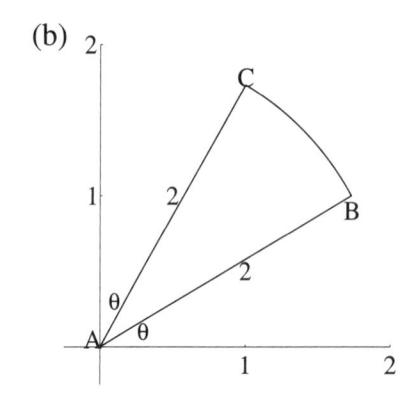

(a)

(b)

3. Draw the following curves.

(a) $x = 3\cos(\theta)$, $y = 2\sin(\theta)$, $-\frac{\pi}{2} \leq \theta \leq \frac{\pi}{2}$,

(b) $x = e^t$, $y = t^2$, $0 \leq t \leq 1$,

(c) $x = te^t$, $y = te^{-t}$, $-1 \leq t \leq 1$.

PROBLEM 15.2
PLANETARY ORBITS

In Worked Example 15.2 it was shown, based on Newton's Laws of Motion and Universal Gravitation, that the planets more along a parametric path given by

$$r(\theta) = \frac{d}{1 + e\cos(\theta)}.$$

Show that if $e<1$ this is an ellipse but if $e>1$ then it is a hyperbola.

(Hint: See the Worked Example 14.2.)

4. Spherical polar coordinates

In plane polar coordinates, the position of an object can be specified by its Cartesian coordinates x and y, or its polar co-ordinates r and θ. Similarly, in three dimensional motion we can use the Cartesian coordinates x, y, and z or its spherical polar coordinates ρ, θ, and ϕ.

Parameterize a path on the surface of a spherical apple of radius 5 cm which will enable an automatic peeler to remove the skin in one strip approximately 0.5 cm wide.

5. Velocity and acceleration in spherical polar coordinates

Express the spherical polar coordinates in terms of polar coordinates involving the usual unit vectors **i**, **j**, and **k** in the directions of the Cartesian axes.

Find expressions for velocity and acceleration in spherical polar coordinates.

Chapter 16 Mathematical Modeling

Prerequisites

This chapter brings together many of the concepts presented in Volumes 1 and 2 of *Calculus Connections*. However, the following list of modules is particularly relevant.

(1) Modules on differentiation (3 and 5).

(2) Modules on integration (6, 7 and 9).

(3 The solution of differential equations (Module 13).

Objectives

Mathematical modeling is the art of taking a physical situation, constructing a mathematical representation, and drawing conclusions from the model in order to make predictions. Many real-life situations are very involved and lead to complicated mathematical models. However, there is little point in constructing elegant and sophisticated models if we cannot solve the mathematical equations that result. In this chapter we look at the process of how mathematical models are constructed; in particular, we examine what is called a *modeling cycle*. The idea of the modeling cycle is to identify the major features, formulate relationships between them, and solve the resulting equations. A very important step is to compare the solution with real life. If the results of the mathematical model agree with reality, then we can consider the model to be validated; but the chances are that we will have to reexamine the original features and repeat this cycle until we get a suitable model.

Connections

It is difficult to talk about modeling in abstract terms and so we shall look at a variety of different mathematical models. What they all have in common is the essential part that the Calculus plays in their development.

The first application is about time, but we shall see that the equation it produces leads to a more wide-ranging discussion of oscillating phenomenas. For many years the swinging of a pendulum was the basis of all clock mechanisms. We shall consider the construction of a mathematical model for the swinging of a pendulum. What are the important features and how are they related? We are interested mainly in the period of the pendulum, that is, the time to complete one complete oscillation so which factors will influence the period? If we change the length of the pendulum, will the period change? What if we change the mass at the end of the pendulum or the material from which it is made? If we move the pendulum to any other location, will it affect the period? All these factors needed to be taken into consideration in the construction of early clocks.

The second application involves a much more up to date phenomenon; central heating, or rather the insulation we install to reduce heating bills. Many factors influence the design of both household and industrial heating systems, and we will try to model a small part of such a system in an attempt to decide the best policy when installing insulation. If we install extensive insulation, then we will cut down heat losses, but to do so will be very expensive. If we install insufficient insulation, then the loss of heat will be expensive. Is it possible to find an optimal thickness?

The Modeling Cycle for a Simple Pendulum

In order to explore the actual process of constructing a mathematical model, we will discuss one particular model. We will return to the same problem in one of the Worked Examples and give more specific details.

In general, constructing a mathematical model consists of the following steps that make up what we shall call a modeling cycle.

(1) *Identifying the Features*

We establish the main variables and parameters and try to decide which are the most important. There is little point in trying to use a model that is too complicated to solve, so ideally we want to keep it as simple as possible. On the other hand, if the model is too simplistic, then it will be likely be unable to deal with the situation we are trying to model. The aim is to eliminate all but the most important features and yet retain a sufficient number to be able to construct a realistic model.

Example 16.1
When trying to model a simple pendulum we need to look at any and all features that might influence the motion. For example:
  (a) the length of the pendulum,
  (b) the mass of the pendulum bob,
  (c) the external forces acting on the pendulum,
  (d) the shape of the pendulum,
  (e) the method of suspension,
  (f) the type of pivot used,
  (g) air resistance.

To include all these features should give a comprehensive model, but it is likely to be very complicated so we try to identify any features that are of less importance. For example, if we assume that the bob is a small, heavy object, we can ignore feature (d). Next, if we assume that the method of suspension is a light, inelastic string suspended from a frictionless pivot, we can ignore (e) and (f). Finally, if we try to include air resistance, (g), which turns out to be related to the square of the speed of the pendulum, we obtain a complicated nonlinear differential equation. However, if we assume that the pendulum is only moving slowly, we can effectively ignore this term. We are then left with a model that involves only the features (a), (b), and (c).

(2) *Formulation*

Next we establish how the important features are related by mathematical equations.

Example 16.2
We need to find a mathematical relationship between the features identified in Example 16.1. Let the length of the pendulum be l, (m), and assume that its mass is m, (kg); then the external forces acting on the pendulum bob are

(i) the tension in the pendulum's string, T,

(ii) the force exerted by the pull of the earth, P,

both measured in Newtons.

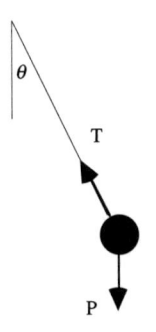

Figure 16.1. The forces on a simple pendulum bob.

Now the angle that the pendulum makes to the vertical to be θ; as shown in Figure 16.1, then the force perpendicular to the string is given by $-P\sin(\theta)$. Using Newton's Laws of Motion, we obtain

$$ml\theta'' = -P\sin(\theta)$$

and since the pull of the earth, P, is given by mg where g is the acceleration due to gravity,

$$l\theta'' = -g\sin(\theta).$$

(See Chapter 15 for a derivation of acceleration in polar coordinates.) This is the equation we must solve to investigate the motion of the pendulum.

(3) Solution

We take any equations that result from the formulation stage and try to solve them. If they are too complicated then we attempt to simplify them within the restrictions imposed in the formulation stage.

Example 16.3

In Example 16.2 we produced a model for the motion of a pendulum which gave the equation

$$l\theta'' = -g\sin(\theta).$$

This is a nonlinear second order differential equation. However, it is not possible to solve this equation analytically. Although we could solve it numerically, we will try another approach. For small angles θ we can approximate $\sin(\theta) \approx \theta$, in which case the above differential equation becomes

$$\theta'' = -\omega^2\theta, \quad \omega^2 = \frac{g}{l}.$$

This is a linear second order differential equation. We saw in Chapter 13 that the solution of this equation is given by

$$\theta(t) = A\cos(\omega t - \phi),$$

which is an oscillatory solution.

(4) Interpretation and Validation

This is the hard part. If a model is to be of any use, it must not only fit any assumption made in the formulation stage but must also be able to make predictions about the way the real situation behaves. We can then compare the predictions with the real world. If they agree, we have a good model. If they do not agree, we need to go back to the formulation step and see whether any assumptions can be made to give a more realistic model, and then we repeat the whole process.

Example 16.4

The solution of the differential equation derived in Example 16.3,

$$\theta'' = -\omega^2\theta, \quad \omega^2 = \frac{g}{l},$$

is

$$\theta(t) = A\cos(\omega t - \phi),$$

which will continue to oscillate with the same amplitude. However, a real pendulum slows down, and to include this effect we would need to revise the model. You will get an opportunity to investigate this in Exercise 16.1.

> ### Exercise 16.1
>
> **Validating the pendulum model**
> In Example 16.3 we produced a model for the motion of a pendulum which resulted in the differential equation
> $$\theta'' = -\omega^2\theta, \quad \omega^2 = \frac{g}{l}.$$
> The general solution of this equation is
> $$\theta(t) = A\cos(\omega t - \phi).$$
> This model predicts that the period of the pendulum, T, is given by $\dfrac{2\pi}{\omega} = 2\pi\sqrt{\dfrac{l}{g}}$ and that
>
> (a) is independent of the mass of the pendulum,
> (b) is proportional to $\sqrt{l}$.
>
> Run the application **Pendulum** in the software for this module and confirm these predictions.

Building a Model

To see the process of building a model in action, we will examine another situation that is an everyday phenomenon. When a car hits a bump in the road, it bounces up and down due to the suspension, but the damping produced by the shock absorbers will eventually cause the oscillations to die away. (See Figure 16.2.) Let us try to construct a model for this.

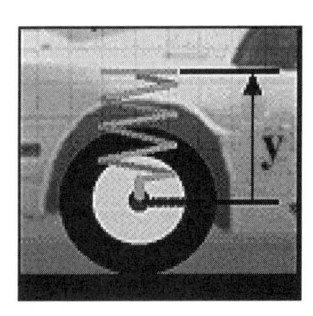

Figure 16.2. Simple car suspension.

Example 16.5
At each wheel there is a large spring, or some equivalent mechanism, which is designed to cushion the occupants when the car's wheels strike any unevenness. To model all the springs would be very complicated, so let us assume that each wheel acts independently and so that we have only a single spring to look at. Other features include any damping in the system produced by the shock absorbers, air resistance, and driving conditions. These are going to make life difficult, so let's ignore them and try to develop a simple model based only on the suspension.

If we ignore all other features, then the only forces acting in the vertical direction are gravity and the restoring force whenever the spring is compressed. The next step is to assign variables:

m the mass of the vehicle (kg),
y the distance of the spring mounting above the axle (m),
l the natural length of the spring (m),
r the radius of the wheel (m),
g the acceleration due to gravity (m/sec^2),
T the tension in the spring (Newtons).

Formulation

Now we have to establish how the features listed above are related

> ### Exercise 16.2
>
> **Compression of a spring**
> Use the software to simulate a simple spring and show that the tension in a compressed/extended spring is proportional to the amount it is compressed/extended.

The restoring force, T, in a spring is proportional to the amount of its compression or extension. If the spring is compressed to a length y, then the restoring force can be written as

$$T = k(l - y),$$

where k is a constant which depends on the materials from which the spring is made.

The only forces acting on the car are gravity, mg, and the tension in the spring, T. From Newton's Second Law of Motion[1],

$$my'' = T - mg$$

or

$$my'' = k(l - y) - mg \quad (16.1)$$

Notice that if the car is stationary, then the height of the car above the spring mounting is given by

$$y = l - \frac{mg}{k} \quad (16.2)$$

Exercise 16.3

Compression of a spring

Equation (16.2) suggests that if the car is stationary, then its suspension is compressed by $\frac{mg}{k}$; that is, the suspension of a stationary car is directly proportional to the total mass of the car. How would you set about checking whether this conclusion is valid?

Solving the Equations

Equation (16.1) can be written as

$$y'' + \omega^2 y = C \quad (16.3)$$

where

---
[1] For simplicity we can take Newton's Second Law as: mass times acceleration equals net forces.

$$\omega^2 = \frac{k}{m}, \quad C = \frac{kl}{m} - g.$$

The general solution of the homogeneous equation

$$y'' + \omega^2 y = 0$$

is given in Module 13 as

$$y(t) = A \cos(\omega t - \phi).$$

A particular solution of the inhomogeneous equation is $y_p(t) = K$, where K is a constant, since the right hand side of equation (16.3) is a constant. This gives

$$K = l - \frac{mg}{k}$$

which is the compression of the spring when it is stationary. Let us put

$$y_0 = l - \frac{mg}{k},$$

then the solution of the differential equation (16.1) is given by

$$y(t) = y_0 + A\cos(\omega t - \phi).$$

Exercise 16.4

Interpretation of the solution

Plot the function

$$y(t) = y_0 + A\cos(\omega t - \phi),$$

$$y_0 = l - \frac{mg}{k}, \quad \omega = \sqrt{\frac{k}{m}},$$

and investigate the effect of the parameters l, m, k, and arbitrary constants A and ϕ.

The solution of the differential equation involves two arbitrary parameters A and ϕ. However, if the car hits an object of height h, which compresses the suspension, then, before the car itself begins to rise

$$y(0) = l - h, \quad y'(0) = 0.$$

However,

$$y(0) = A\cos(\phi) + y_0, \quad y'(0) = \omega A \sin(\phi).$$

The second equation gives $\phi = 0$, and the first, $A = l - h - y_0$. Substituting for y_0 gives the solution for the differential equation (16.1) as

$$y(t) = y_0 + (\frac{mg}{k} - h)\cos(\omega t - \phi), \quad \omega = \sqrt{\frac{k}{m}}.$$

Interpretation and Validation

What does the solution above represent?

Exercise 16.5

Interpreting the solution
Plot the solution

$$y(t) = y_0 + (\frac{mg}{k} - h)\cos(\omega t - \phi),$$

$$y_0 = l - \frac{mg}{k}, \quad \omega = \sqrt{\frac{k}{m}},$$

for different values of the parameters l, m, and k and initial displacement h.

Does this represent the motion of a car after the suspension has been compressed?

Exercise 16.5 predicts that the motion of the automobile is an undamped oscillation about the steady-state position of the car. However, in the real world the damping in the suspension of the car will eventually reduce the oscillation. Therefore, we return to the original feature list and see what additional information we can include to improve the model.

Refining the Model

An important part of a car's suspension is provided by the shock absorbers, (see Figure 16.3), which provide the damping missing from the model.

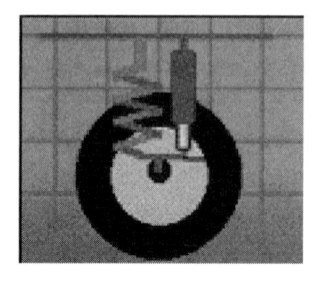

Figure 16.3. Inclusion of shock absorbers.

The damping provided by the shock absorbers will oppose any motion; the faster the motion, the greater the damping forces.

Exercise 16.6

Refining the model - damping
Investigate the relationship between the stiffness of the shock absorber, speed of the motion, and the damping forces exerted.
Use the software to show that the damping effect is proportional to the velocity y', that is, $D = ry'$.

The addition of damping will change the equation of motion, (16.1), to the inhomogeneous second order differential equation

$$my'' = k(l - y) - ry' + mg \quad (16.4)$$

The solution of this differential equation is given in Chapter 13. We begin by looking at the solution of the homogeneous equation

$$my'' = -ky - ry' \quad (16.5)$$

which is given by

$$y(t) = Ae^{\alpha t},$$

where α is a solution of the quadratic equation

$$m\alpha^2 + r\alpha + k = 0 \quad (16.6)$$

This equation has two solutions given by

$$\alpha = \frac{-r \pm \sqrt{r^2 - 4mk}}{2m}.$$

If $r^2 - 4mk > 0$, then both solutions are negative and the solution of the differential equation is

$$y(t) = A_1 e^{\alpha_1 t} + A_2 e^{\alpha_2 t}, \quad \alpha_1, \alpha_2 < 0.$$

If $r^2 - 4mk = 0$, then the solutions of the characteristic equation are equal and the solution of the differential equation is

$$y(t) = (A_1 + A_2 t) e^{\alpha_1 t}, \quad \alpha_1 < 0.$$

If $r^2 - 4mk < 0$, then the solution of the characteristic equation is a complex conjugate pair,

$$\alpha = \mu \pm i\omega,$$

with

$$\mu = -\frac{b}{2m}, \quad \omega = -\frac{\sqrt{4mk - r^2}}{2m},$$

and the solution of the differential equation is

$$y(t) = e^{\mu t}(A_1 \cos(\omega t) + A_2 \sin(\omega t)),$$

or

$$y(t) = e^{\mu t} A \cos(\omega t - \phi).$$

Example 16.6

The solution of the differential equation

$$y'' + 2ry' + y = 0, \quad r \neq \pm 1,$$

is of the form $y(t) = Ae^{\alpha t}$ where α is a solution of the equation

$$\alpha^2 + 2r\alpha + 1 = 0.$$

This equation has two solutions given by

$$\alpha = -r \pm \sqrt{r^2 - 1}$$

and the behavior of the solution depends on the value of r. For example, if $r=2$ then $\alpha = -2 \pm \sqrt{3}$ and the solutions of the differential equation, which is a combination of

$$y_1(t) = e^{(-2-\sqrt{3})t}$$

and

$$y_2(t) = e^{(-2+\sqrt{3})t}$$

will decay away, as shown in Figure 16.4.

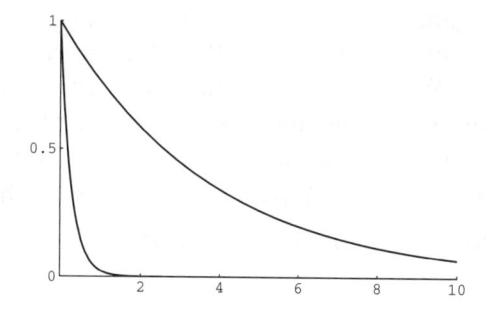

Figure 16.4. Decaying exponential solutions.

Alternatively, if $r=0.5$ then $\alpha = -0.5 \pm \sqrt{-0.75}$ and the solution of the differential equation, which is a combination of

$$y_1(t) = e^{-0.5t} \cos(\sqrt{0.75}\, t)$$

and

$$y_2(t) = e^{-0.5t} \sin(\sqrt{0.75}\, t)$$

will decay, as shown in Figure 16.5.

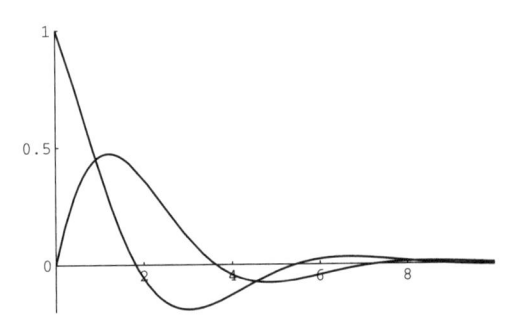

Figure 16.5. Decaying oscillating solutions.

Exercise 16.7

Investigate the solution of the differential equation
$$y'' + 2ry' + y = 0$$
when r=1.

Let us now return to the solution of the inhomogeneous equation

$$my'' = k(l-y) - ry' - mg.$$

We have seen that the solution of the homogeneous equation involves a negative exponential and must therefore decay to zero. We will now consider the effect of the inhomogeneous term. Since this term is a constant, we look for a particular solution $y_p(t) = K$. Substituting this into the above differential equation we get

$$0 = k(l-K) - mg$$

and so

$$K = l - \frac{mg}{k}$$

as in the undamped case. Therefore, the solution of the inhomogeneous differential equation is

$$y(t) = y_0 + \psi(t), \quad y_0 = l - \frac{mg}{k},$$

where $\psi(t)$ is the general solution of the homogeneous equation (16.5). If the solutions of the characteristic equation, (16.6), are complex, $\alpha = \mu \pm i\omega$, then

$$\psi(t) = e^{\mu t} A \cos(\omega t + \varphi),$$

If the solutions of (16.6) are real and distinct, say α_1 and α_2, then

$$\psi(t) = A_1 e^{\alpha_1 t} + A_2 e^{\alpha_2 t}.$$

In the first case $\mu < 0$ and in the second α_1 and α_2 are both negative and so $\psi(t) \to 0$ and the solution decays to steady state solution

$$y_0(t) = l - \frac{mg}{k}.$$

Exercise 16.8

Investigate the solution of
$$my'' = k(l-y) - ry' - mg$$
when the roots of the characteristic equation, (16.6), are equal.

Exercise 16.9

Interpreting the solution
Does the motion predicted have any qualitative agreement with the motion you observe in a real suspension system?

We have now completed two loops of the modeling cycle and in doing so have constructed a mathematical model that has enabled us to investigate a very simple suspension system. In the problems associated with this module, you will be asked to investigate this model further.

WORKED EXAMPLE 16.1
SIMPLE PENDULUM

A swinging pendulum was the basis of clock mechanisms for many years. In this example we will look at a further development of the model for the motion of a simple pendulum.

Features

We have already seen that we can develop a model for the swinging of a pendulum based on:

> length of pendulum l (m),
> angle θ (radians),
> mass of bob m (kg),
> tension in the string T (Newtons),
> gravity g (m/sec^2).

Any other features such as :

> air resistance,
> shape of the bob and pendulum,
> friction at the pivot,
> the weather,
> type of suspension used,

complicate the model so we will just ignore them for the moment.

Formulation

We now try to formulate some mathematical relationships between the features listed above. The first stage is to decide on the coordinate system to use. If we use cartesian coordinates (x,y) then we will have two components to consider. However, if we use polar coordinates (r,θ), then r is fixed and only θ changes with time. This gives a simpler parameterisation. (See Chapter 15.)

From Figure 16.1 the forces acting on the pendulum are

(a) perpendicular to the string: $-mg\sin(\theta)$,

(b) along the string towards the point of suspension: $T - mg\cos(\theta)$.

In Chapter 15, the component of acceleration perpendicular to the string is given by $l\theta''$. Therefore, applying Newton's Second Law of Motion in this direction gives

$$ml\theta'' = -mg\sin(\theta)$$

or

$$\theta'' = -\frac{g}{l}\sin(\theta).$$

This is a second order nonlinear differential equation. However, if we assume that the pendulum only swings through small angles, then $\sin(\theta)$ is approximately equal to θ, and we can approximate the differential equation by

$$\theta'' = -\frac{g}{l}\theta \ .$$

To justify this assumption look at Figure 16.6, where it is shown that, provided the angle θ, (measured in radians), is "small", θ is a good approximation for $\sin(\theta)$.

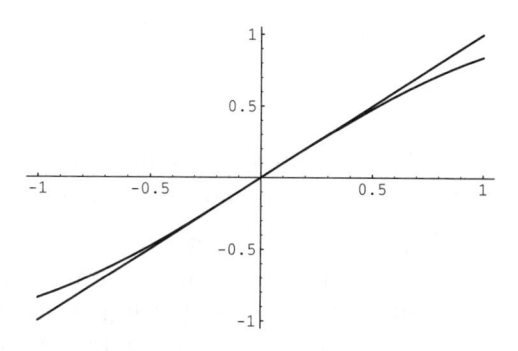

Figure 16.6. A comparison of $\sin(\theta)$ and θ for $|\theta|<1$.

Solving the equation

The general solution of the second order linear differential equation

$$\theta'' = -\frac{g}{l}\theta$$

is given in Chapter 13 as

$$\theta(t) = A\cos(\omega t - \phi)$$

where $\omega = \sqrt{\frac{g}{l}}$, and A and ϕ are arbitrary constants. The constants A and ϕ can be determined if the initial position and speed of the pendulum are given. For example, if the pendulum starts at an angle of 0.1 radian but is initially at rest, then $\theta(0)=0.1$ and $\theta'(0)=0$ which gives $A=0.1$ and $\phi=0$ so that $\theta(t) = 0.1\cos(\omega t)$. This solution is shown in Figure 16.7.

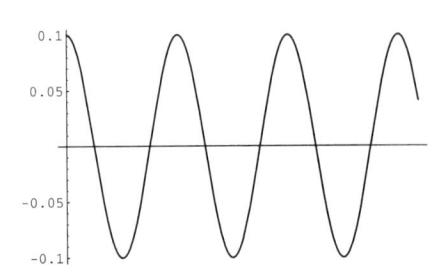

Figure 16.7. Predicted solution $\theta(t) = 0.1\cos(\omega t)$.

Validation

This model for a simple pendulum predicts that the period of the pendulum is independent of the mass but is proportional to $\sqrt{l}$. Exercise 16.1 demonstrated that, provided the pendulum only swings through a small angle, this agrees with a real pendulum. This agreement between the model and reality gives some credence to the model.

Initially, however, a real pendulum swings to and fro with uniform period but as it does so the amplitude dies away. Unfortunately, our model predicts that the pendulum will keep on swinging with the same amplitude. We see that there is some qualitative agreement, but the long term behavior is different. We need to revise the model we have developed to see if we can deal with this discrepancy.

Refining the model - making it better

Clearly, we need to improve our model by returning to the features we ignored. For example, we ignored air resistance, since the pendulum was moving relatively slowly, and friction at the pivot, since this makes the model harder to solve.

The next stage in the modeling process would be to include such terms. To do this let us look again at the equation of motion of the pendulum:

$$ml\theta'' = -mg\sin(\theta).$$

Any frictional forces, F, in the suspension will reduce the acceleration, and so we can revise this equation to give

$$ml\theta'' = -mg\sin(\theta) - F.$$

Furthermore, it is clear that the faster the pendulum is moving the greater the friction, so we shall assume that any frictional force is proportional to θ' that is

$$F = r\theta'$$

where r is a constant. The equation of motion now becomes

$$ml\theta'' + r\theta' + mg\sin(\theta) = 0.$$

Finally, we use the small angle approximation for $\sin(\theta) \approx \theta$ to give

$$ml\theta'' + r\theta' + mg\theta = 0.$$

The solution of this equation is given by

$\theta(t) = Ae^{\lambda t}$ where

$$\lambda = \mu \pm i\omega,$$

are the roots of the equation

$$ml\lambda^2 + r\lambda + mg = 0.$$

Therefore,

$$\mu = -\frac{r}{2ml}, \quad \omega = \frac{\sqrt{4m^2gl - r^2}}{2ml}.$$

Provided, $r^2 < 4m^2gl$, then the solution of the differential equation is oscillatory and is given by

$$\theta(t) = Ae^{\mu t}\cos(\omega t - \varphi).$$

Furthermore, since $\mu < 0$, such solutions will decay to zero as does a real pendulum. Therefore, we have been able to refine the model to predict the decaying solution of the pendulum.

WORKED EXAMPLE 16.2
HEAT LOSSES IN A PIPE

Construct a model to investigate the optimum thickness of insulation needed in a central heating system to minimize the overall cost involved taking into account the heat losses in a section of pipe and the thickness of insulation used.

To establish the overall problem, notice that if we use very little insulation then it does not cost much to install, but we will pay dearly for the heat losses in the pipe. However, if we use a lot of insulation then there will be a lot less loss of heat, but we will have to pay more for the insulation. Somewhere between these extremes we hope to find an optimum solution.

What features will influence the loss of heat?

Heat losses through an element of insulation around a pipe can occur through **radiation, convection,** and **conduction.** Let us assume that the most important of these is the heat lost to conduction. This is just the first of a series of assumptions we will need to make in order to build a model. Other assumptions could include:

(a) Ignore heat losses in the copper pipe itself.

(b) Assume that everything is a steady state; that is, the heating has been on for a sufficient time that everything has settled down and that we need only consider the heat lost through the insulation and not the heat flowing through the pipe.

(c) Assume that the insulation is made up of concentric layers and that the overall cost is annualized, that is, averaged over the expected lifetime of the insulation.

(d) Assume that we are looking at the combined costs over a year during which the heating is on for a specified period based on data gathered over previous years.

Next we assign variables:

length of pipework, L (m)
radius of the pipe, r (m)
radius of outer insulation, R (m)
thickness insulation, t (m)
thermal conductivity of insulation, k
temperature at radius x, $T(x)$ °C
total heat lost, Q (Watts)
annualised cost of insulation, M per m thickness,
cost of power, P per WH
heating period, H hours
water temperature, T_r, °C
room temperature, T_R, °C

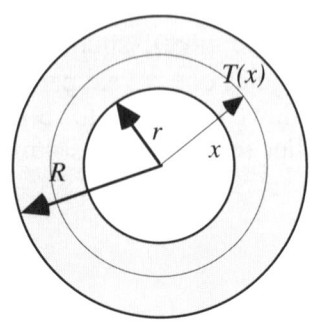

Figure 16.8. Cross section through a pipe of radius R and insulation of external radius R.

Formulation

Next we derive an equation for the rate of loss of heat, q, through a typical element of insulation which is at a distance x from the center of the pipe and has length L as shown in Figure 16.9.

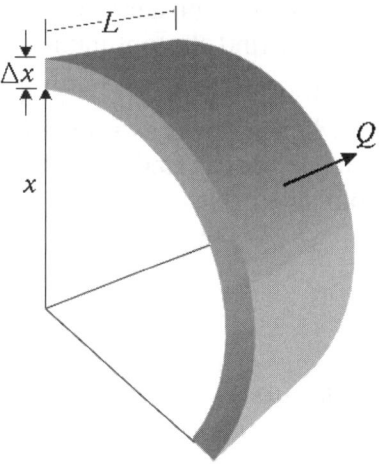

Figure 16.9. A typical element of the insulation material.

The rate of heat loss through this section is proportional to

(i) the cross-sectional area of the element A,

and

(ii) the temperature gradient through the insulation

$$\frac{T(x+\Delta x)-T(x)}{\Delta x}.$$

We now let the thickness of the element in Figure 16.9 tend to zero, in which case

$$\frac{T(x+\Delta x)-T(x)}{\Delta x} \to \frac{dT}{dx}.$$

Therefore, if k is the thermal conductivity of the insulation, then

$$Q = -kA\frac{dT}{dx} \text{ watt/hours}$$

This is called Fourier's law. The total heat lost is then given by QH. Notice that the area of a circular element of radius x and length L is $A=2\pi xL$.

Formulation of equations

In a steady state the total costs are given by adding the annual heating costs, C_H, to the annualized cost of installing the insulation C_L; that is:

Overall Cost/unit length = $C_H + C_L$.

The annualized insulation cost per unit length, C_L, is

(Insulation thickness)(Cost/unit length/year)

so that $\qquad C_L = tM.$

The heating cost/unit length/hour is given by

(Heat Lost/unit length)(Cost /KWH)

so that
$$C_H = \frac{QHP}{L}.$$

Finally, the heat lost through a cylinder of radius x is $Q = -k2\pi xL\frac{dT}{dx}$ so that

$$\frac{dT}{dx} = -\frac{Q}{k2\pi xL}.$$

This is a first order differential equation for T as a function of x, that we need to solve in order to find an expression for Q.

Solution

Recall that we have assumed everything is in a steady state so that Q is constant. Integrating this differential equation gives

$$T(x) = -\frac{Q}{k2\pi L}\ln(x) + T_0,$$

where T_0 is a constant of integration. However, we know that $T(r)=T_r$ so that

$$T_0 = T_r + \frac{Q}{k2\pi L}\ln(r)$$

and

$$T(x) = T_r - \frac{Q}{k2\pi L}(\ln(x) - \ln(r)).$$

Finally, since $T(R)=T_R$ we can solve for Q to get

$$Q = \frac{2k\pi L(T_R - T_r)}{\ln(R) - \ln(r)}.$$

where $R=r+t$. Therefore, the total cost of heating and insulation is given by

$$C(t) = \frac{2k\pi(T_R - T_r)}{\ln(t+r) - \ln(r)}PH + Mt.$$

We wish to minimize this as a function of t. Differentiating with respect to t gives

$$C'(t) = M - \frac{2k\pi PH(T_R - T_r)}{(r+t)(\ln(r+t) - \ln(r))^2}.$$

At a turning point $C'(t)=0$; therefore,

$$(r+t)\ln\left(\frac{r+t}{r}\right)^2 = \frac{2k\pi(T_R - T_r)PH}{M}$$

which turns out to be a minimum. We need to solve this equation to find the optimum value of t. To do this let us write $z = 1 + \frac{t}{r}$, so that $r+t=zr$, and then the above equation becomes

$$z\ln(z)^2 = K \qquad (16.8)$$

with

$$K = \frac{2k\pi(T_R - T_r)PH}{Mr}.$$

Notice that K is a constant for the system, Therefore, to find the optimum thickness of insulation we plot $y(z) = z\ln(z)^2$ and observe when $y(z)=K$, as shown in Figure 16.10. Hence, we can find the optimal thickness from $t=r(z-1)$.

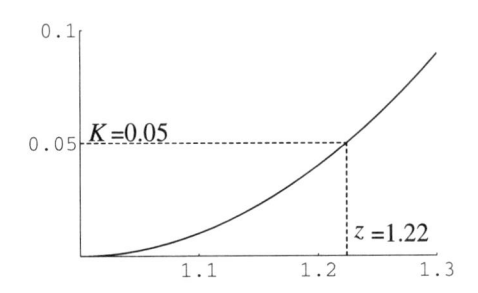

Figure 16.10. Solution of equation (16.8).

Validation

How would you validate this model? Can you think of an experiment that would validate it?

Refining the model

How would you include the cost of installing any insulation? How about including the thickness of the copper pipe itself? Alternatively, we could consider the cost of borrowing the money to pay for the installation of the insulation?

Can you think of any other features that might improve this model?

PROBLEM 16.1
TRAFFIC CALMING

Name: _____

Date: _____

Section: _____

Many towns and cities attempt to control the speed of local traffic using ramps in the road. The ramps are placed across the road at intervals so that any vehicle traveling at excessive speed will bounce uncomfortably at best and can be damaged. Construct a mathematical model that will restrict traffic to less than 15 Km/hour based on ramps D meters apart as shown in Figure 16.11

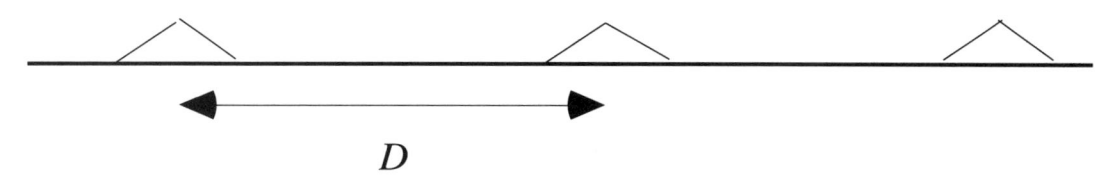

Figure 16.11. Traffic calming ramps.

(Hint:

Among the many ways of approaching this problem, you might like to consider:

(a) Assume that the driver accelerates uniformly for as long as possible and then applies the brakes so that the car arrives at the next ramp at a "suitable speed",

or

(b) Relate the distance between the ramps and design the suspension of the car so that the oscillations induced as the car passes over one ramp have not decayed by the time the next ramp is reached.)

PROBLEM 16.2
KITCHEN SCALES

Name: _____

Date: _____

Section: _____

Many simple scales are based on a spring mechanism as shown in Figure 16.12. As objects are placed on the scale pan, the spring is extended and a needle records the extension of the spring, which in turn indicates the weight of the object. However, placing an object on the scale makes the pan vibrate, Adapt the model of the car suspension to model the behavior of a set of scales.

Suggest suitable characteristics for a set of scales for weighing objects between 10 gm and 4 kg where the scales are sufficiently sensitive to be able to detect a change of 10 gm and yet sufficiently damped that when an object is placed on the scales any oscillation dies away within two periods.

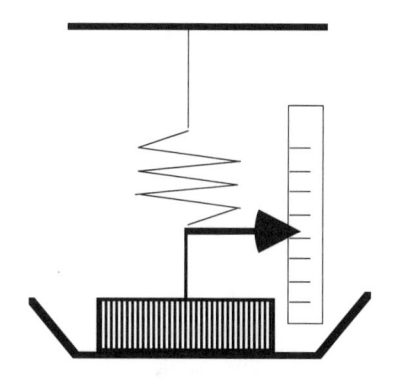

Figure 16.12. Simple Scales.

NEW SITUATIONS

1. Planetary motion

By patient observation, over many years Johannes Kepler recorded the position of the planets in our solar system and deduced that they move in ellipses around the sun and do so in a very precise way. He formulated three 'laws':

(a) The orbits of the planets are ellipses having the sun at one focus.

(b) A line joining a planet to the sun sweeps out equal areas in equal time periods.

(c) The square of the period of revolution of a planet is proportional to the cube of its mean distance to the sun.

However, Kepler was unable to provide the mathematics behind his conclusions. This was left to Sir Isaac Newton who was able to show that, based on relatively simple assumptions, the paths of the planets could be determined very precisely.

Newton's inspiration is called the Law of Universal Gravitation. This law states that any two bodies attract each other with a force that is proportional to their masses and inversely proportional to the square of the distance between them.

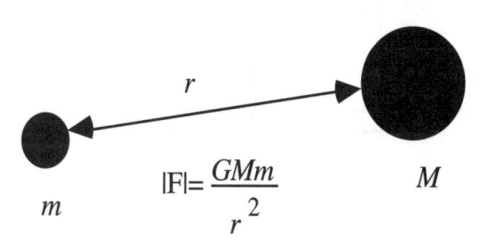

Figure 16.13. The Law of Universal Gravitation.

In addition Newton's Laws of Motion give that if an object of mass m is travelling so that its displacement is given by the vector $\mathbf{r}$ and it is subject to a force $\mathbf{F}$ then

$$m\mathbf{r}'' = \mathbf{F}.$$

In Worked Example 15.2 it was shown that a consequence of the Law of Universal Gravitation is that the force $\mathbf{F}$ only has a component in the direction of the radial vector $\mathbf{r}$ so that the acceleration is given by

$$\mathbf{r}'' = \left(r'' - r(\theta')^2\right)\mathbf{e}_r + \frac{1}{r}\frac{d}{dt}\left(r^2\theta'\right)\mathbf{e}_\theta.$$

The radial component of this acceleration is then

$$m\left(r'' - r(\theta')^2\right) = -\frac{GMm}{r^2}$$

and the transverse component is

$$m\frac{1}{r}\frac{d}{dt}\left(r^2\theta'\right) = 0.$$

This last equation gives

$$r^2\theta' = C \quad \Rightarrow \quad \theta' = \frac{C}{r^2},$$

where C is an arbitrary constant. Substituting into the radial equation gives

$$\frac{d^2r}{dt^2} - \left(\frac{C}{r^2}\right)^2 r = -\frac{GM}{r^2}$$

which is a second order non-linear ordinary differential equation for r as a function of the parameter t. In Chapter 15 we saw that if we now make a change of variable

$$r(t) = \frac{1}{u(\theta)}$$

then

$$\frac{d^2u}{dt^2} + u = \frac{GM}{C^2}.$$

This differential equation has the solution

$$u(\theta) = A\cos(\theta + \phi) + B,$$

where A and ϕ are arbitrary constants and

$$B = \frac{GM}{C^2}$$

which is a fixed constant for any planetary system. Returning to the original variable r and parameter t, we get

$$r(\theta) = \frac{1}{u(\theta)} = \frac{1}{A\cos(\theta + \phi) + B}.$$

which represents an elliptic orbit

$$r(\theta) = \frac{d}{1 + e\cos(\theta)}.$$

We can now use Kepler's laws to validate this model.

(a) Clearly, the orbital path predicted using Newton's Laws of Motion and Gravitation confirms Kepler's first law.

(b) The area swept out by a radius vector r as it moves through an angle $\Delta\theta$ is given by

$$\tfrac{1}{2}r^2\,\Delta\theta.$$

However, the assumption that the force of attraction only has a component in the direction along r leads to

$$m\frac{1}{r}\frac{d}{dt}\left(r^2\theta'\right) = 0$$

which is validated by Kepler's second law.

How would you use Kepler's third law to provide further validation?

2. Hot coffee?

When making coffee, some people put the milk in first and then add hot coffee, others put the hot coffee in first and then add the milk. Assuming that the milk comes straight from the refrigerator and the coffee is almost boiling, which method will keep the coffee hottest assuming a two minute delay between the two liquids being mixed?

(Newton's law of cooling of a liquid predicts that heat is lost at a rate proportional to the difference between the temperature of the liquid and the surrounding air.)

3. Bacterial growth

A biologist is investigating the growth of a colony of bacteria by observing the area they occupy in a laboratory specimen dish and records the following data.

Time t (hrs.)	Area H (sq. cm.)
1	0.025
2	0.180
3	1.192
4	5.001
5	8.808
6	9.820
7	9.975
8	9.997

Plot these data and suggest which of the following differential equations would be suitable to model the data.

(1) $\dfrac{dH}{dt} = aH + b$

(2) $\dfrac{dH}{dt} = aH(H+b)$

(3) $\dfrac{dH}{dt} = \dfrac{H+a}{H+b}$

(4) $\dfrac{dH}{dt} = aH^2 + b$

Suggest suitable values for the parameters a and b. In practice, the area occupied by the bacteria approaches an upper limit. Explain whether your selected model has this behavior and predict the maximum area using your model.

4. Carbon dating

A historian wishes to date some artefacts using carbon dating. To do this, she measures the percentage of Carbon 14 in the sample and compares it with the percentage in a comparable live specimen. If the half life of Carbon 14 is approximately 5000 years and the artefact has 60% of the Carbon 14 of the live sample, estimate its age.

(The law of radioactive decay states that a radioactive element of mass m decays at a rate proportional to its mass. The half-life of a radioactive element is the time for a sample of mass m to decay to half this mass.)

Program Documentation

System Requirements
Windows 3.1 or higher (including Windows 95), 486DX microprocessor, 8MB RAM, 10MB free hard disk, Multispin CD-ROM drive, SVGA graphics card and monitor (256 colors), 100% compatible Soundblaster™ sound card.

Installation Instructions
1. Insert CD-ROM into CD drive
2. In the Windows Program Manager, click *Run* under the *File* menu
3. Click *Browse*
4. Choose the CD-ROM drive by clicking in the lower right corner
5. Click *install.exe*
6. Click *OK* twice
7. Follow the instructions on screen

Navigation
When the program has installed, you'll see 3 icons: *Main Menu, Intro, Readme*. Refer to the Readme file for any installation or performance problems. Click on Intro for a 2-minute Introduction to the philosophy of the program. Click on Main Menu to get to the content of the program.

Main Menu
The main menu lists the titles of the 8 modules in this volume. Click on the topic you want to access.

Module Menu
Our goal is to make the software as user-friendly as possible, so each module is set up exactly the same way. There are Applications (move the cursor over the Applications icon to see list, and then click on the topic you want to explore), Concepts (move the cursor over the concepts icon to see list, and then click on the topic you want to explore), and Exercises (ditto). From this module menu you can access anything in the module, in any order you choose.

Start Connections
There's also a prescribed pathway through the Applications, Concepts and Exercises in each module. Bring the cursor up from the bottom middle of the screen, upwards toward the middle until "Start Connections" pops up. Click on "Start Connections" to enter this prescribed mode. This pathway takes about 45 minutes to 1 hour to complete so it's perfect for one lab session. This suggested route makes it effortless for instructors to make assignments, or for students to always know where in the program they are, and where to go next.

Navigation

Navigational Icons and Tools
Navigational icons are located on the bottom right of each screen. Use them to:
- *Go to next screen*
- *Return to previous screen*
- *Rewind to beginning of section*
- *Fast forward to next section*
- *Return to module menu*

Other navigational icons are presented when appropriate. Use them to:
- *Repeat voiceover or animation*
- *Give additional information, problem statements or instructions to run a simulation*

Other Icons and Tools
Hotwords
Words highlighted in a contrasting color are "hot." Click on them for definitions and other information.

Calculus Connections provides convenient on-line Help, Options, Tools and References. Bring up this menu by passing the cursor over the bottom of the screen, and click to select.

Help
On-line interactive Help is always available. Click Help for information on:
- *Entering equations*
- *Syntax errors*
- *Interacting with plots*
- *Main module menu*
- *Menu bar navigation*
- *Menu bar icons*
- *Menu bar pull downs*

Options
Click to change setup, colors, sounds and Preferences.

Mathematical Tools
Click on Maple, Mathematica, Derive or Mathcad for access if they are resident on your computer.

Calculus Connections provides custom 2-D and 3-D Graphing Tools. You can plot, manipulate and save graphs of functions without complicated syntax or other software.

References
Click on References for definitions and biographical information.

Exit
Exit lets you quickly access the Main Menu, Module Menu, or Introduction to Calculus Connections. You can also quit the Program from the Exit menu.

Volumes 1 & 3

Technical Assistance
Call **212-850-6753**, or send email to **math@jwiley.com** for assistance.

Calculus Connections Volumes 1 & 3

Volume 1
- Lines, Functions, and Equations
- Limits
- Rates of Change and Differentiation
- Transcendental and Inverse Functions
- Applied Maximums and Minimums
- Areas as Limits
- Fundamental Theorem of Calculus
- Mean-Value Theorem for Integrals

Volume 3 *(available spring 1997)*
- Scalar Functions of More than One Variable
- Vector Valued Functions
- Directional Derivatives and the Gradients of Functions of One Variable
- Double and Triple Integrals
- Triple Integrals in Cylindrical and Spherical Coordinates
- Centroids, Center of Gravity and Moments of Inertia
- Line Integrals
- Surface Integrals

To Order Volumes 1 & 3

To purchase a copy of Calculus Connections Volume 1 & 3, visit your favorite bookstore or call Wiley directly at **1-800-225-5945**.

Wiley will pay shipping and handling costs for all pre-paid orders.

partial differential equation 72
partial sum 51
phase planes 82
piecewise parametric 117
planetary motion 149
planetary orbits 127
polar coordinates 95,96,114
population growth 71,73
power series 57

Queuing 5,15

Radial speed 119
range 95
rectilinear motion 21
reduction formulas 19
rpm 36
rules of integration 10,12

Separation of variables 78
sequences 52
series 52
series, bounded 54

series, decreasing 54
series, increasing 54
series, oscillatory 53
series, geometric 53
SHM, Simple Harmonic Motion 79
simple pendulum 132,139
Simpson's Rule 35
Simpson's rule, error 40
single-valued 98
spherical polar coordinates 102
standard integrals 9
sum rule for integration 10

Taylor series 51,58
terminal velocity 28
transverse speed 119
trapezoidal rule 36
trapezoidal rule, error 49

Unit vectors 120

Velocity, in polar coordinates 120

STAFFORD LIBRARY
COLUMBIA COLLEGE
1001 ROGERS STREET
COLUMBIA, MO 65216